ELECTROMAGNETISM

Titles in the *Foundations of Engineering* Series

FOUNDATIONS OF ENGINEERING
Series Editor: G. E. Drabble

Electromagnetism
R. G. Powell

MACMILLAN

First published 1990

Published by
MACMILLAN EDUCATION LTD
Houndmills, Basingstoke, Hampshire RG21 2XS
and London
Companies and representatives
through the world

Typeset by P & R Typesetters Ltd, Salisbury, Wiltshire, UK
Printed in Hong Kong

British Library Cataloguing in Publication Data
Powell, R.G.
  Electromagnetism
  1. Electromagnetism
  I. Title   II. Series
  537

  ISBN 0-333-48317-0

Series Standing Order
If you would like to receive future titles in this series as they are
published, you can make use of our standing order facility. To place a
standing order please contact your bookseller or, in case of difficulty,
write to us at the address below with your name and address and the
name of the series. Please state with which title you wish to begin your
standing order. (If you live outside the United Kingdom we may not
have the rights for your area, in which case we will forward your order
to the publisher concerned.)

Customer Services Department, Macmillan Distribution Ltd
Houndmills, Basingstoke, Hampshire RG21 2XS, England.

To Janice, Charis and Jason

CONTENTS

SERIES EDITOR'S FOREWORD

This series of programmed texts has been written specifically for first year students on degree courses in engineering. Each book covers one of the core subjects required by electrical, mechanical, civil or general engineering students, and the contents have been designed to match the first year requirements of most universities and polytechnics.

The layout of the texts is based on that of the well-known text, *Engineering Mathematics* by K. Stroud (first published by Macmillan in 1970, and now in its third edition). The remarkable success of this book owes much to the skill of its author, but it also shows that students greatly appreciate a book which aims primarily to help them to learn their chosen subjects at their own pace. The authors of this present series acknowledge their debt to Mr Stroud, and hope that by adapting his style and methods to their own subjects they have produced equally helpful and popular texts.

Before publication of each text the comments of a class of first year students, of some recent engineering graduates and of some lecturers in the field have been obtained. These helped to identify any points which were particularly difficult or obscure to the average reader or which were technically inaccurate or misleading. Subsequent revisions have eliminated the difficulties which were highlighted at this stage, but it is likely that, despite these efforts, a few may have passed unnoticed. For this the authors and publishers apologise, and would welcome criticisms and suggestions from readers.

Readers should bear in mind that mastering any engineering subject requires considerable effort. The aim of these texts is to present the material as simply as possible and in a way which enables students to learn at their own pace, to gain confidence and to check their understanding. The responsibility for learning is, however, still very much their own.

G. E. Drabble

HOW TO USE THIS BOOK

Electromagnetism is an essential part of a curriculum in electrical and electronic engineering, and a highly desirable part of any broadly based programme in engineering or applied science. It is fundamental to the understanding of the design and application of the whole range of electrical and electronic devices and equipment from the tiny integrated circuits encountered in electronics to the very large machines associated with the power supply system. Without such understanding, the reasons for the differences between 'real' components and the ideals of circuit theory will remain forever a mystery.

In the manner of K. Stroud's *Engineering Mathematics* mentioned in the Series Editor's Foreword, this book consists of a number of programmes (nine in all) each of which contains a number of frames. The frames are usually less than one page in length. Most of them end with a question of some sort, the answer to which is given in a box at the beginning of the following frame. For best results, always try to work out the answer *before* referring to the next frame.

At regular intervals throughout each programme there are revision exercises in the form of a set of 'true or false' questions to test your understanding of the fundamentals, and a number of numerical problems to test whether you can apply the theory. As these are designed to consolidate the work covered up to that point, it is *NOT* a good idea to skip them. The answers to the numerical problems are given in the last frame of each programme.

There are three occasions when I have resorted to the irritating phrase 'it can be shown that . . .', which often leads readers to suspect that the author can't do it himself! However, I believe that in these instances, there would be nothing to be gained in presenting the proofs (which are somewhat long-winded) nor any virtue in asking you to look them up for yourself.

ACKNOWLEDGEMENTS

I am most grateful to several anonymous reviewers for their helpful and constructive criticism and suggestions. I am also indebted to the many authors whose books I have consulted over the years. In particular I found those of Professors P. Hammond and G. Carter to be stimulating and infectious in their enthusiasm for the subject.

My thanks are also due to Dr Peter Holmes for encouraging me to write this book and to the many students with whom I have had the pleasure of working at Nottingham (formerly Trent) Polytechnic and who, by their questioning, have helped me to see things more clearly.

Nottingham, October 1989   R. G. POWELL

Programme 1

REVISION AND BASIC
BACKGROUND INFORMATION

1

INTRODUCTION

This programme sets out the basic background information which will be required in later programmes.

We begin by considering the importance of units and dimensions in engineering and how a system of units is formulated. Having obtained an answer to a problem, you should always ask yourself 'is this a reasonable magnitude for this particular quantity?', and this means acquiring a thorough knowledge of the units in which various quantities are measured. Does one tesla represent a strong magnetic field? Is it reasonable to expect forces of several hundred newtons to exist in electrostatic field systems? Can five hundred million watts of power be generated in a single electrical machine? When you have worked through the programmes in the book, you should be able to answer such questions, which you may not be able to answer now.

The difference between scalar and vector quantities is then explained and the Principle of Superposition is defined. The use of this principle often results in vector quantities having to be added, and techniques for doing this are given.

We then survey the mathematics required for the programmes in the book. The treatment of electromagnetism followed in this text concentrates on its physical aspects and its uses, the mathematical aspects being kept to a minimum. Nevertheless, there will be times when you will have to differentiate or integrate some functions, manipulate algebraic equations or trigonometrical relationships and calculate the areas and volumes of some common geometrical shapes.

Although some topics, such as solid angles, line integrals, surface integrals and volume integrals may be new to you, much of the mathematics required will be familiar, but of course, you should revise where necessary.

The programme ends with a brief introduction to alternating quantities.

When you have studied this programme, you should be able to:

- derive the dimensions of any new quantity you meet
- test equations for dimensional balance
- express any multiple or sub-multiple of any unit in terms of any other multiple or sub-multiple of that unit
- add vector quantities
- differentiate and integrate simple functions
- determine turning points and decide whether they are maxima or minima
- determine the areas and volumes of common geometric shapes
- state the solid angle subtended at a point by a small plane area
- represent a point in space in cartesian, polar or complex form
- represent a sinusoidally alternating quantity by means of a phasor diagram.

2

DIMENSIONS AND UNITS

In all branches of engineering, measurement is essential and this involves two concepts which are interconnected. To begin with we must decide what it is we need to measure, a length perhaps or a force or an electrical potential. We must then give this quantity a unit to describe its magnitude: how long is the line? how strong is the force? how big is the potential difference?

In electromagnetism four basic quantities are required and all other quantities can be derived from these.

The four basic quantities, called *dimensions*, are mass, length, time and electric current. The dimensions of all other quantities are derived from these, either directly (for example, velocity = length/time) or by statements of proportionality (for example, force $\propto$ mass $\times$ acceleration). Different systems of units define these basic quantities in different ways and assign different values to the various constants of proportionality. Fortunately, a coherent system of units is now in general use and is called the 'Système International d'Unités', abbreviated to the 'SI'.

Now see if you can state the dimensions of acceleration.

3

$$\boxed{\text{length}/(\text{time})^2}$$

... because acceleration = velocity/time = (length/time)/time.

It is usual to write these dimensional equations in square brackets as follows (the basic dimensions being in capital letters):

$$[a] = [v][T^{-1}] = [LT^{-1}][T^{-1}] = [LT^{-2}]$$

The SI basic unit of mass is the kilogramme (kg) and is defined as the mass of an actual piece of metal (platinum–iridium) kept under controlled conditions at the international bureau of weights and measures in Paris.

The SI basic unit of length is the metre (m) and is defined in terms of a number of wavelengths of the radiation arising from the transition between certain energy levels of the krypton-86 atom.

The definitions of the other two basic units are given in the next frame.

4

The SI basic unit of time is the second (s) and is expressed in terms of a multiple of the period of the radiation of the caesium-133 atom.

The SI basic unit of electric current is the ampere (A) and is defined as that current which, when maintained in each of two infinitely long, straight, parallel conductors of negligible cross-sectional area, placed 1m apart in a vacuum, produces a mutual force of 2×10^{-7} newtons per metre length between them.

Two other basic units (not relevant to the programmes in this book) are required for the complete specification of the SI. These are chosen to be the degree Kelvin (K) for thermodynamics and the candela for illumination.

Given that electric current is equivalent to the amount of electric charge per second passing a certain reference point, determine the dimensions of electric charge.

5

$$\boxed{[TA]}$$

Since current $[A]$ = charge $[q]$/time $[T]$, then $[q] = [T][A] = [TA]$.

DIMENSIONAL ANALYSIS

Every equation should be dimensionally balanced and this is a necessary condition for the correctness of an equation. It is often helpful to check equations using dimensional analysis to see if they are balanced. It is also possible to determine the dimensions of a quantity using this technique.

Example

The magnitude of the force between two charges is given by $F = q_1 q_2 / 4\pi\varepsilon d^2$ where q_1 and q_2 are the two charges, d is the distance between them and ε is a constant.

Remembering that charge = current × time, determine the dimensions of ε.

The solution is given in the next frame. Try it yourself, first.

$$[M^{-1}L^{-3}T^4A^2]$$

Here is the working: first we transpose the equation to make ε the subject so that $\varepsilon = \dfrac{q_1 q_2}{4\pi F d^2}$. Next, we write down the dimensions of each of the quantities on the right-hand side of the equation, thus

$$[q] = [TA]$$

4π is a pure number and has no units. It is said to be dimensionless.

$$[F] = [M][a] = [M][v][T^{-1}] = [M][LT^{-1}][T^{-1}] = [MLT^{-2}]$$

$$[d^2] = [L^2]$$

Finally we write down the dimensional equation for ε:

$$[\varepsilon] = \frac{[TA][TA]}{[MLT^{-2}][L^2]} = [TA][TA][M^{-1}L^{-1}T^2][L^{-2}]$$

$$= [M^{-1}L^{-3}T^4A^2]$$

Which of the following are units:
torque, second, newton, time, kilogramme?

$$\boxed{\text{second, newton, kilogramme}}$$

There is an enormous range of magnitudes in the quantities encountered in electromagnetism. For example, electric currents can be smaller than 0.000 000 001 ampere and bigger than 1000 ampere. Rather than having to write many noughts in these cases, multiples and sub-multiples of units are used. The common ones are given below.

Multiple/sub-multiple	Abbreviation	Value
giga	G	10^9
mega	M	10^6
kilo	k	10^3
milli	m	10^{-3}
micro	μ	10^{-6}
nano	n	10^{-9}
pico	p	10^{-12}

8

As you can see from the table, the preferred multiples and sub-multiples are ten to the power ± 3. These preferred multiples should normally be used. For convenience, the centimetre (10^{-2} m) is often used for length.

Note that the multiples are abbreviated to a capital letter whereas the sub-multiples are abbreviated to a small (lower case) letter. The exception is the kilo which is abbreviated to k to avoid confusion with degrees Kelvin (K).

Example

Express 1 A (one ampere): (a) in milliamperes, (b) in kiloamperes.

Solution

(a) To find how many milliamperes in one ampere, divide by 10^{-3}.

$$\therefore 1 \text{ A} = 1/10^{-3} \text{ mA} = 1000 \text{ mA} = 10^3 \text{ mA}$$

(b) Similarly, to find the number of kA in 1 A, divide by 10^3.

$$\therefore 1 \text{ A} = 1/10^3 \text{ kA} = 0.001 \text{ kA} = 10^{-3} \text{ kA}$$

Now you try these:

Express: (i) 10 mA in kA
         (ii) 1.5 km in mm
        (iii) 1 μF in pF
       (iv) 500 MW in GW.

9

(i) 10^{-5} kA	(ii) 1.5×10^6 mm
(iii) 10^6 pF	(iv) 0.5 GW

A physical quantity may be either a *scalar quantity* or a *vector quantity*. The essential difference is that, whereas a scalar quantity is defined simply by a given number of units, a vector quantity must be assigned a direction as well as a number for complete specification.

Which of the following quantities are vectors:

area, mass, force, time, velocity, speed, acceleration?

> force, velocity, acceleration

Vector quantities throughout this book are given in bold italic type (for example, F).
Two vectors are equal if they have the same magnitude and direction so that the
statement $F_1 = F_2$ means that the forces F_1 and F_2 have the same magnitude and
direction. The statement $F_1 = 0.5F_2$ means that the force F_1 has the same direction
as that of F_2 and is half as big as it. What do you think is the meaning of the statement
$F_1 = -F_2$?

> F_1 is a vector whose magnitude is the same as that of
> F_2 and whose direction is opposite to that of F_2

Unlike scalar quantities, for which addition is straightforward, when adding vector
quantities, account must be taken of their directions. There are a number of techniques
available for doing this and these are best illustrated by means of examples.

Example

Three forces act at a point P as shown in the diagram (i) below. Determine the total
force at the point P.

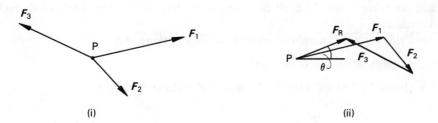

(i)   (ii)

Solution 1

If the three vectors are drawn (the order is unimportant) in a chain as shown in the
diagram (ii) above, then the line joining the start of the first vector F_1 to the end of
the last vector F_3 is the sum of the three vectors and is called the resultant vector
F_R. The direction of F_R is E $\theta°$ N.

12

Solution 2

This method involves completing the parallelogram of which the sides are pairs of vectors. For example, we take the forces F_1 and F_2 and, by completing the parallelogram abcd, determine their resultant F_{12} which is the diagonal of the parallelogram as shown in diagram (i) below.

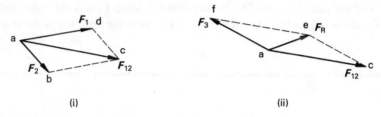

(i)   (ii)

Next we complete the parallelogram acef for which the sides are F_{12} and F_3. The diagonal of this parallelogram is the resultant (F_R) of the three forces and this is shown in diagram (ii) above.

13

Solution 3

This method requires each vector to be resolved into two components at right angles. The steps in the method are:

(i)   choose two directions at right angles
(ii) resolve each vector into its components along these directions
(iii) add the components in each direction separately to find the resultant in each direction
(iv) use Pythagoras's theorem to obtain the sum of these two.

For the problem under consideration

(i)   We choose (arbitrarily) the horizontal and vertical directions:

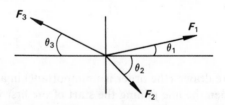

(ii) The horizontal components of the three forces are $F_1 \cos \theta_1$ acting due east; $F_2 \cos \theta_2$ acting due east; $F_3 \cos \theta_3$ acting due west: the vertical components of the forces are $F_1 \sin \theta_1$ acting due north; $F_2 \sin \theta_2$ acting due south; $F_3 \sin \theta_3$ acting due north.

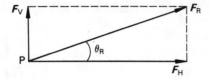

(iii) The resultant horizontal force is given by

$$F_H = F_1 \cos \theta_1 + F_2 \cos \theta_2 - F_3 \cos \theta_3 \text{ acting due east}$$

The resultant vertical force is given by

$$F_v = F_1 \sin \theta_1 - F_2 \sin \theta_2 + F_3 \sin \theta_3 \text{ acting due north}$$

Write down an expression for the resultant force F_R.

14

$$\boxed{\begin{array}{c} F_R = (F_H^2 + F_v^2)^{1/2} \text{ acting} \\ \text{as shown with } \theta_R = \tan^{-1}(F_v/F_H) \end{array}}$$

THE PRINCIPLE OF SUPERPOSITION

The need to add vectors often comes about as a result of applying the Principle of Superposition to a problem. This Principle is one which may be applied to many different types of engineering problem and whose general statement is as follows: 'in any linear system, the total response at any instant is the vector sum at that instant of the responses to all stimuli applied to the system, each individual response being dependent upon its own associated stimulus and independent of all other stimuli and responses'.

Although the general statement mentions the *vector* sum, there are cases where a simple algebraic sum applies. An example is in direct current electric circuits. If a number of batteries (stimuli) are connected in such a circuit, the current (response) in any circuit element may be found by determining the current in that element when each battery acts alone and then adding the results. An example of the use of the Principle of Superposition follows in Frame 15.

15

Example

Diagram (i) shows two forces acting simultaneously at a point P. Determine the magnitude and direction of the resultant force at P.

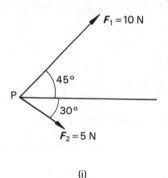

(i)   (ii)

Solution

Diagram (ii) shows the forces resolved into their horizontal and vertical directions. Their resultant at P is F_R directed as shown and is obtained by adding F_1 and F_2 in accordance with the Principle of Superposition.

The component of F_1 in the horizontal direction $= 10 \cos 45° = 7.07$ N due east.
The component of F_2 in the horizontal direction $= 5 \cos 30° = 4.33$ N due east.
The resultant horizontal force $(F_H) = 7.07 + 4.33 = 11.4$ N due east.
The component of F_1 in the vertical direction $= 10 \sin 45° = 7.07$ N due north.
The component of F_2 in the vertical direction $= 5 \sin 30° = 2.5$ N due south.
The resultant vertical component $(F_v) = 7.07 - 2.5 = 4.57$ N due north.
The resultant total force at P $F_R = ((11.4)^2 + (4.57)^2)^{1/2} = 12.28$ N at an angle $\phi = \tan^{-1}(4.57/11.4) = 21.8°$, i.e., in a direction E 21.8° N.

16

Now decide which of the following statements are true:

(a) 'Quantity' is another name for 'Unit'.
(b) The dimensions of all electromagnetic quantities may be derived from those of four basic quantities.
(c) The SI unit of mass is the kilogramme (kg).
(d) Velocity is a scalar quantity.
(e) The SI unit of length is the centimetre because of its convenient size.
(f) 10 GW = 10 000 MW.
(g) 100 nF = 0.1 pF.
(h) The Principle of Superposition was developed for adding vectors.
(i) Vector quantities can only be added graphically.
(j) Two vectors are equal only if they have the same magnitude and direction.

17

True: b, c, f, j

(a) Units are the means by which quantities are assigned magnitude.
(d) Speed is a scalar quantity; velocity has direction and so is a vector.
(e) The SI unit is the metre (Frame 3).
(g) 100 nF = 100 000 pF.
(h) The Principle of Superposition applies to any linear system (Frame 14).
(i) Analytical methods are often preferred.

18

SUMMARY OF FRAMES 1–17

There are six basic SI units from which all the other units in the system are derived. These are the units of mass (kilogramme), length (metre), time (second), electric current (ampere), temperature (Kelvin) and luminous intensity (candela).

Scalar quantities are specified by a unit and a number whereas vector quantities require in addition a statement of direction for complete specification.

The resultant effect of a number of forces (or other stimuli) acting simultaneously may be determined using the Principle of Superposition.

19

EXERCISES

(i) Determine the dimensions of: (a) force (mass × acceleration); (b) work (force × distance); (c) power (work per unit time); (d) potential (work/charge). See Frame 5 for the dimensions of charge.

(ii) Express: (a) 3.2 cm in mm; (b) 6.2 MW in kW; (c) 108 μJ in mJ; (d) 0.5 μF in pF; (e) 3.3 kV in mV; (f) 300 mA in μA.

(iii) Two forces act at a point as follows: $F_1 = 15$ N acting due west. $F_2 = 25$ N acting in a direction W 30° N.
Determine the resultant force at the point.

20

As stated in Frame 1, we will be concentrating on the physical aspects of electromagnetism in this book. However mathematics often makes statements easier to express and arguments and proofs easier to develop. The following 23 frames preview the mathematics which you will encounter in later programmes.

TRIGONOMETRICAL RATIOS

The angles of a right-angled triangle may be expressed as ratios of the lengths of two of its sides. These ratios are defined as follows:

The *sine* (abbreviated to sin) of an angle is the ratio of the side opposite the angle to the hypothenuse of the triangle so that $\sin \phi = c/a$.

The *cosine* (cos) of an angle is the ratio of the side adjacent to the angle to the hypothenuse of the triangle so that $\cos \phi = b/a$.

The *tangent* (tan) of an angle is the ratio of the side opposite the angle to the side adjacent to the angle so that $\tan \phi = c/b$.

Write down expressions for $\sin \theta$, $\cos \theta$ and $\tan \theta$ in the above triangle.

21

$$\boxed{\sin \theta = b/a; \cos \theta = c/a; \tan \theta = b/c}$$

Here are some more definitions:

The reciprocal of the sine of an angle is called the cosecant (cosec).
The reciprocal of the cosine of an angle is called the secant (sec).
The reciprocal of the tangent of an angle is called the cotangent (cot.)

Referring to the triangle in the previous frame, give expressions for:
$\sec \phi$, $\operatorname{cosec} \theta$, $\cot \phi$.

$$\sec \phi = a/b; \; \operatorname{cosec} \theta = a/b; \; \cot \phi = b/c$$

TRIGONOMETRIC IDENTITIES

Referring to the diagram in Frame 20, we see that

$$\sin \theta = b/a \qquad \qquad \text{(i)}$$
and
$$\cos \theta = c/a \qquad \qquad \text{(ii)}$$

Dividing (i) by (ii), we have $\dfrac{\sin \theta}{\cos \theta} = \dfrac{b/a}{c/a} = \dfrac{b}{c} = \tan \theta$. The relationship $\dfrac{\sin \theta}{\cos \theta} = \tan \theta$ is true for all values of θ and is called an *identity*.

Again, $(\sin \theta)^2$ (which is more usually written $\sin^2 \theta$) $= b^2/a^2$
and $\cos^2 \theta = c^2/a^2$.

It follows that $\sin^2 \theta + \cos^2 \theta = \dfrac{b^2 + c^2}{a^2} = \dfrac{a^2}{a^2} = 1$

i.e. $\qquad \qquad \sin^2 \theta + \cos^2 \theta = 1 \qquad \qquad$ (iii)

There are very many more identities relating the six functions (sin, cos, tan, sec, cosec and cot) and their derivation and application may be found in many textbooks on mathematics.

There is one further identity which is of use to us and it concerns a compound angle $(A + B)$. It is given here without derivation:

$$\cos(A + B) = \cos A \cos B - \sin A \sin B$$

It follows that $\cos 2\theta \; (= \cos(\theta + \theta)) = \cos \theta \cos \theta - \sin \theta \sin \theta$

i.e. $\qquad \qquad \cos 2\theta = \cos^2 \theta - \sin^2 \theta$

$$= 1 - \sin^2 \theta - \sin^2 \theta \text{ (making use of (iii) above)}$$

$$= 1 - 2 \sin^2 \theta.$$

Rearranging, we obtain $\sin^2 \theta = (1 - \cos 2\theta)/2$.
We will find this identity to be very useful in a later frame in this programme.

23

RECTANGULAR AND POLAR COORDINATES

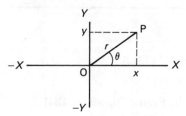

The point P shown in the diagram may be represented by (x, y) which means 'x units along the X-axis and y units along the Y-axis'. x and y are called *rectangular* or *cartesian coordinates*.

The point P can also be represented by (r, θ) where r is the distance (OP) from the origin and θ is the angle between the positive X-axis and the line OP. r and θ are called *polar coordinates*.

Express the polar coordinates r and θ in terms of the rectangular coordinates x and y.

24

$$r = (x^2 + y^2)^{1/2}; \; \theta = \tan^{-1}(y/x)$$

Note that the point P in the previous frame is located in the so-called first quadrant (i.e. both x and y are positive). It could just as well have been located at a $(-x, y)$ in the second quadrant, at $(-x, -y)$ in the third quadrant or at $(x, -y)$ in the fourth quadrant.

It is often convenient to express rectangular coordinates in complex form, using the so-called j-notation. In this form, the x term, measured along the horizontal axis is said to be 'real'. The letter j is placed before the y term (mathematicians use the letter i *after* the term) which is called the quadrature or the imaginary term and is measured along the vertical axis. In complex number theory j (or i) has the value $\sqrt{(-1)}$ which is imaginary. In this book you need simply regard it as a label indicating that the number which follows is measured along the vertical axis.

Points P and Q are located at $(3, 4)$ and $(-3, 4)$ respectively. Express point P in complex form and point Q in polar form.

$$3 + j4; 5 \angle 126.9°$$

MENSURATION

As well as plane surfaces, we will frequently encounter cylinders and spheres in later programmes. It would be useful, therefore, to memorise the formulae for the surface areas and volumes of these shapes.

For a cylinder of radius r and length l:

> the cross-sectional area is πr^2 (m^2)
> the curved surface area is $2\pi rl$ (m^2)
> the total surface area is $2\pi r^2 + 2\pi rl = 2\pi r(r + l)$ (m^2)
> the volume is $\pi r^2 l$ (m^3)

For a sphere of radius a:

> the surface area is $4\pi r^2$ (m^2)
> the volume is $(4/3)\pi r^3$ (m^3)

Example

The volume of the space between two concentric spheres is 920 cm^3. The radius of the outer sphere is 7 cm. Calculate the radius of the inner sphere and the surface area of the outer ~~space~~ *sphere*.

Solution

Volume of the outer sphere $= (4/3)\pi(7)^3 = 1437$ cm^3.
∴ volume of the inner sphere $= 1437 - 920 = 517$ cm^3.
Let the radius of the inner sphere be r, then $(4/3)\pi r^3 = 517$
    so that $r^3 = 123.4$ cm^3 and $r = \underline{4.98 \text{ cm}}$.
The surface area of the outer sphere $= 4\pi(7)^2 = \underline{617.75 \text{ cm}^2}$.

Now try this one.
A cylindrical wire is 5 m long and its diameter is 0.5 mm. Calculate its cross-sectional area, its curved surface area and its volume.

26

$$\boxed{0.196 \text{ mm}^2, \ 7854 \text{ mm}^2, \ 980 \text{ mm}^3}$$

... the cross sectional area is $\pi(0.25)^2$ mm^2
the curved surface area is $2\pi(0.25) \times 5000$ mm^2
the volume is the cross-sectional area (mm^2) $\times$ the length (mm).

SOLID ANGLES

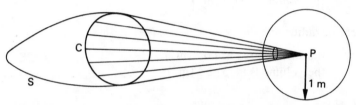

If straight lines are drawn to the point P from points on the contour C of the open surface S shown in the diagram, a cone is formed with its apex at P. (An open surface is one which has an edge or contour; a closed surface is one which has no edge. A good example is an egg which is a closed surface until the top is cut off when it becomes an open surface.)

The contour C is said to subtend a solid angle at P. A solid angle is an angle in the three-dimensional sense and its unit is the steradian.

To obtain a measure of the solid angle, we draw a sphere of unit radius with P as centre and determine the area intercepted on this sphere by the cone of lines as they make their way to P. If the area intercepted is 1 m^2, then the solid angle subtended at P by the contour C is 1 steradian. In other words, a unit solid angle is produced if a unit area is intercepted on a sphere of unit radius. The area intercepted may be of any shape.

The areas of similar figures increase as the square of their linear dimensions so that the area intercepted by the cone of lines on a sphere of radius r will be r^2 times as great as that intercepted on the unit sphere. Suppose that the area intercepted on the sphere of radius r by the cone of lines from C is A. What, then, is the solid angle subtended at P?

$$\boxed{A/r^2 \text{ steradians}}$$

If the area intercepted on a sphere of radius r is A, then the area intercepted on a sphere of radius 1 (which is the definition of the solid angle at P) is A/r^2.

THE SOLID ANGLE SUBTENDED BY A CLOSED SURFACE AT A POINT INSIDE IT

If lines are drawn from all points on the closed surface to a point inside it, they must all intercept the whole of the surface area of a unit sphere drawn about the point. The surface area of a unit sphere is $4\pi(1)^2$ so that the solid angle subtended at the point is 4π steradians.

This would be the case if the point P in the diagram in the previous frame were inside the surface S and the contour C were shrunk to zero (i.e. if the surface were closed).

What would be the solid angle subtended at P in the previous diagram if P were in the plane of the contour C?

$$\boxed{2\pi \text{ steradians}}$$

... in this case the cone of lines would intercept half of the sphere of unit radius drawn about P.

THE SOLID ANGLE SUBTENDED BY A CLOSED SURFACE AT A POINT OUTSIDE IT

Consider again the diagram in Frame 26 and with point P fixed, imagine the contour C to shrink. As it does so the solid angle which it subtends at P will become smaller and eventually, when the contour disappears completely, the solid angle which it subtends at P will become zero.

Summarising:

the solid angle subtended by a closed surface at a point inside it $= 4\pi$ units
the solid angle subtended by an open surface at a point in its plane $= 2\pi$ units
the solid angle subtended by a closed surface at a point outside it $= 0$.

29

THE SOLID ANGLE SUBTENDED AT A POINT BY A SMALL PLANE AREA

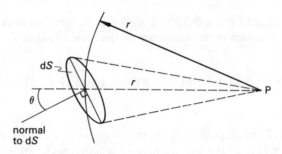

The lateral dimensions of dS (a small plane area of any shape whatever) are small compared with its distance (r) from the point P.

A spherical surface of radius r is drawn with P as its centre.

A cone of lines drawn from the contour of dS to the point P will intercept on the sphere an area d$S \cos \theta$.

What, therefore, is the solid angle subtended at P by the small plane area dS?

30

$$\boxed{(\mathrm{d}S \cos \theta)/r^2 \text{ steradians}}$$

… because the area intercepted on a sphere of unit radius is $1/r^2$ times as great as that intercepted on the sphere of radius r.

DIFFERENTIATION

The standard differential coefficients which you will be required to use are given here. It would be useful, though not essential, to memorise them.

$y = f(x)$:	x^n	e^{mx}	$\ln x$	$\sin mx$	$\cos mx$	$\tan mx$	$\cot mx$
$\mathrm{d}y/\mathrm{d}x$:	nx^{n-1}	me^{mx}	$1/x$	$m \cos mx$	$-m \sin mx$	$m \sec^2 mx$	$-m \operatorname{cosec}^2 mx$

Write down the differential coefficients of (i) x^7, (ii) $a^{1/2}$, (iii) m^{-3}, (iv) $b^{-1/2}$, (v) $\sin 3x$, (vi) $\tan y$, (vii) $e^{x/2}$.

> (i) $7x^6$; (ii) $(1/2)a^{-1/2}$; (iii) $-3m^{-4}$; (iv) $(-1/2)b^{-3/2}$;
> (v) $3\cos 3x$; (vi) $\sec^2 y$; (vii) $(1/2)e^{x/2}$

DIFFERENTIATION OF A PRODUCT

If $y = uv$ where u and v are both functions of x, then

$$\frac{dy}{dx} = u\frac{dv}{dx} + v\frac{du}{dx}$$

Read this as 'the first times the differential of the second plus the second times the differential of the first'.

Example

If $y = 4x^3 \ln x$, then using the above formula with $u = 4x^3$ and $v = \ln x$:

$$\frac{dy}{dx} = 4x^3(1/x) + (\ln x)(12x^2) = 4x^2(1 + 3\ln x)$$

Now you try this one: if $y = x^4 \sin 2x$, find dy/dx.

> $(x^4)(2\cos 2x) + (\sin 2x)(4x^3)$

DIFFERENTIATION OF A QUOTIENT

If $y = u/v$ where u and v are both functions of x, then

$$\frac{dy}{dx} = \frac{v(du/dx) - u(dv/dx)}{v^2}$$

Read this as 'the denominator times the differential of the numerator minus the numerator times the differential of the denominator all over the denominator squared'.

For example, if $y = \dfrac{\sin 5x}{4x^2}$, find $\dfrac{dy}{dx}$.

The solution is given in the next frame. Try it yourself first.

33

Solution

$$\frac{dy}{dx} = \frac{(4x^2)(5\cos 5x) - (\sin 5x)(8x)}{(4x^2)^2} = \frac{4x(5x\cos 5x - 2\sin 5x)}{16x^4}$$

$$= \frac{(5x\cos 5x - 2\sin 5x)}{4x^3}$$

34

MAXIMUM AND MINIMUM TURNING POINTS

When we differentiate the equation of a graph, the result is an expression for the gradient of the graph at every point. Thus, for a straight line graph whose general equation is $y = mx + c$, $dy/dx = m$ and this is the gradient of the line at every point.

Again, for a curve represented by $y = 2x^2 + 3x + 2$, $dy/dx = 4x + 3$ and from this the gradient of the curve at any value of x can be found.

Now consider the curve shown in diagram (i) below.

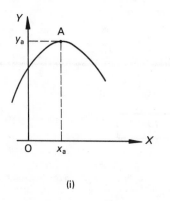

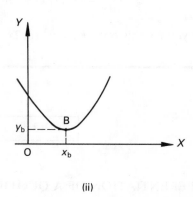

(i)   (ii)

Point A is called *a maximum turning point* and notice that the gradient near A changes from being positive for $x < x_a$ through zero at x_a to negative for $x > x_a$. The value y_a is called a maximum value of y ($= f(x)$).

For the curve shown in diagram (ii), point B is called a *minimum turning point* and the gradient near B changes from being negative for $x < x_b$ through zero at x_b to positive for $x > x_b$. The value y_b is called a negative value of y ($= f(x)$).

For a maximum or minimum turning point, therefore, dy/dx (the gradient) at the point must be zero.

Is it true also that, if $dy/dx = 0$, then there must be a maximum or a minimum turning point?

> No, because the gradient of a curve could be positive
> on both sides of a point where $dy/dx = 0$ (or negative on
> both sides of such a point).

These points are called *points of inflexion* and examples (C and D) are given in the following diagram.

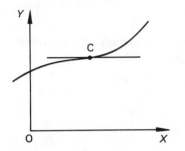 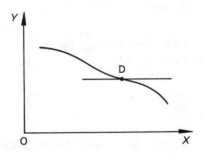

To determine whether the graph of a function (of x, say) has a maximum or a minimum turning point, therefore, we first of all differentiate its equation with respect to x and equate the result to zero. If there is a solution to this equation, then, at the corresponding value of x,

... there is a maximum turning point if dy/dx changes from $+$ to 0 to $-$

... there is a minimum turning point if dy/dx changes from $-$ to 0 to $+$

... there is a point of inflexion if dy/dx changes from $+$ to 0 to $+$ OR from $-$ to 0 to $-$

Note that, as the graph passes through point A (diagram (i) in the previous frame), dy/dx changes from positive to negative. It decreases as x increases. The rate at which dy/dx increases is therefore negative, which may be stated mathematically thus: $\dfrac{d}{dx}\left(\dfrac{dy}{dx}\right) < 0$ and is abbreviated to $\dfrac{d^2y}{dx^2} < 0$.

Note also that, as the graph in diagram (ii) passes through point B, dy/dx changes from negative to positive, i.e. it increases as x increases so that we may write $\dfrac{d^2y}{dx^2} > 0$.

Having found a value for x where dy/dx is zero, therefore, we can differentiate a second time; if this gives a negative result there is a positive turning point, whereas if the result is positive there is a minimum turning point. What if the result is zero?

36

$$\boxed{\text{If } d^2y/dx^2 = 0 \text{ there could be a point of inflexion}}$$

However, d^2y/dx^2 can also be zero at maximum and minimum turning points so it is better in this case to use the first method described in Frame 35.

Example

Determine whether the graph of $y = 4x^2 - 6x + 3$ has a maximum or a minimum turning point.

Solution

$dy/dx = 8x - 6$ and this is zero when $x = 6/8 = 0.75$.

Using method 1, we note that at $x = 0.7$, $dy/dx = 5.6 - 6 = -0.4$.
and that $x = 0.8$, $dy/dx = 6.4 - 6 = +0.4$.
Since dy/dx changes from negative through zero to positive there is a minimum turning point.
  Using method 2, we differentiate a second time.
  $d^2y/dx^2 = 8$ which is positive and indicates a minimum turning point.

37

INTEGRATION

The standard integrals which will be needed in later programmes are given below. Again, it would be useful to memorise them.

$f(x)$:	$x^n \ (n \neq -1)$	m/x	e^{mx}	$\sin mx$	$\cos mx$
$\int f(x)\, dx$:	$\dfrac{x^{n+1}}{n+1} + C$	$m \ln(x) + C$	$\dfrac{e^{mx}}{m} + C$	$-\dfrac{\cos mx}{m} + C$	$\dfrac{\sin mx}{m} + C$

Try these: integrate (i) $3x^2$, (ii) e^{5x}, (iii) $3/x$, (iv) $\sin 5x$, (v) $\cos(x/2)$.

> (i) $x^3 + C$;　(ii) $\dfrac{e^{5x}}{5} + C$;　(iii) $3 \ln(x) + C$
>
> (iv) $\dfrac{-\cos 5x}{5} + C$;　(v) $2 \sin(x/2) + C$

LINE INTEGRALS

Line integration is a simple extension of the familiar process of integration and will be encountered frequently in later programmes. A brief introduction is given here.

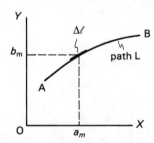

In the diagram, L is a continuous path between the two points A and B. Let $f(x, y)$ be a function of x and y which has a value at every point on L. The path L is divided into n sections, one of which (Δl) is shown. a_m and b_m are the coordinates of one of the points on the section Δl. We now evaluate $f(x, y)$ at every point on the path and multiply by the length of the path over which it has that value (i.e. we form the product $f(a_m, b_m) \times \Delta l$ for each section of the path). On adding these products for all the sections we obtain:

$$\sum_{m=1}^{n} f(a_m, b_m) \times \Delta l$$

where $\displaystyle\sum_{m=1}^{n}$ means 'the sum of the products for all values of m from 1 to n'.

As n becomes infinitely large such that $\Delta l \to 0$, the limit of this sum is known as a line integral and is written

$$\int_{L} f(a_m, b_m) \, dl$$

The L indicates the path over which the integration is to be performed.

39

You will find it helpful to think of ordinary definite integrals as line integrals in which the curve AB (in the diagram in the previous frame) is the X-axis and the integrand is a function of x alone. The following example shows how the evaluation of line integrals can be reduced to the evaluation of ordinary definite integrals.

Example

Evaluate the line integral of $(x + y)\,dx$ over the path which is a part of the graph $y = x + 2$, beginning at $(0, 2)$ and ending at $(4, 6)$.

Solution

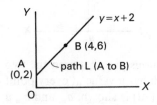

The integrand in line integrals must always be evaluated along the path of integration so we must express y in terms of x. Along $y = x + 2$ this gives us the ordinary definite integral

$$\int_L (x + y)\,dx = \int_0^4 (x + x + 2)\,dx = [x^2 + 2x]_0^4 = (16 + 8) - 0$$

$$= 24$$

All the line integrals which you will have to perform in the following programmes will be quite straightforward and will usually mean adding up all the elemental lengths which make up a line of known length or the circumference of a circle of known radius.

40

SURFACE INTEGRALS

The concept of a line integral may be extended to a surface integral as follows. Let S be a surface over which $f(x, y, z)$ is a continuous function of x, y and z. S is subdivided into n elemental surface areas Δs and a point P_m on Δs has coordinates (a_m, b_m, c_m).

If $f(x, y, z)$ is evaluated at all points such as P_m on all the elemental surfaces Δs which make up the surface S, we can form the product $f(a_m, b_m, c_m) \times \Delta s$.

Write down an expression for the sum of all these products over the whole surface S.

41

$$\boxed{\sum_{m=1}^{n} f(a_m, b_m, c_m) \times \Delta s}$$

The limit of this sum as $n \to \infty$ and $\Delta s \to 0$ is called a surface integral and is written $\iint_S f(a_m, b_m, c_m)\, dS$.

42

VOLUME INTEGRALS

Similarly, for a given function $f(x, y, z)$ and a region of space of volume V, $f(x, y, z)$ can be evaluated at a point P_m (a_m, b_m, c_m) in each of the n elemental volumes (Δv) which make up the volume V.

If the product $f(a_m, b_m, c_m) \times \Delta v$ is formed for each point and these are summed from $m = 1$ to n, the limit of this sum as $n \to \infty$ and $\Delta v \to 0$ is called a volume integral. Remembering how line and surface integrals are written, how do you think a volume integral is written?

43

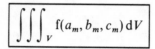

The V indicates the volume over which the integration is to be performed.

The surface and volume integrals which you will have to evaluate in the following programmes will involve simply adding the elemental areas or volumes of well known figures, such as cylinders and spheres.

Now decide which of the following statements are true.

(a) If $\cos \theta = b/a$, then $\operatorname{cosec} \theta = a/b$.
(b) The contangent of an angle is the reciprocal of its tangent.
(c) $\sin \theta = \cos \theta \tan \theta$.
(d) Rectangular coordinates can only be used for points in the first quadrant.
(e) The point at $(1, 1)$ may also be represented by $1.414 \angle 45°$.
(f) The volume of a sphere of radius r is given by $4\pi r^3$.
(g) The surface area of a cylinder of radius r and length l is given by $2\pi r l$.
(h) The unit of solid angle is the steradian.
(i) If $y = e^x$, then $dy/dx = e^x$.
(j) If $y = 1/x$, then the integral of y with respect to x is $(\ln x + C)$.
(k) The graph of $y = f(x)$ has a maximum turning point when $d^2 y/dx^2 > 0$.
(l) Line integrals can be evaluated over curved lines only.

44

True: b, c, e, h, i, j

(a) The reciprocal of $\cos \theta$ is $\sec \theta$.   (Frame 21)
(d) They can be used in any quadrant.   (Frame 24)
(f) The volume is $4\pi r^3/3$.   (Frame 25)
(g) This is the area of the *curved* surface only.   (Frame 25)
(k) It must be <0.   (Frame 35)
(l) Line integrals can also be evaluated over straight lines.   (Frame 39)

45

SUMMARY OF FRAMES 20–44

There are six trigonometric ratios associated with a right-angled triangle. These are sine, cosine, tangent, cosecant, secant and cotangent.

A point in a plane may be represented by rectangular or polar coordinates.

The volume and surface area of a sphere can be expressed in terms of its radius.

The volume and surface area of a cylinder can be expressed in terms of its radius and length.

A solid angle is a three-dimensional angle and is measured in steradians.

It is possible to determine whether a graph has any turning points by differentiating its equation and equating the result to zero.

Line integration is a simple extension of ordinary definite integration.

46

EXERCISES

(i)   A right-angled triangle has sides of length 3 cm, 4 cm and 5 cm. Write down the sine, secant and tangent of its smallest angle.

(ii) A point P is located in a direction W 30° N from a point O. Give the polar and the rectangular coordinates of P, using O as the origin and taking the easterly direction as that of the positive x-axis. The distance between O and P is 5 cm.

(iii) Differentiate the following functions with respect to x:

(a) x^5; (b) $\dfrac{1}{x^4}$; (c) $x(3\sin 2x)$; (d) 6; (e) $\dfrac{e^x}{\sin x}$; (f) $x \ln x$

(iv) Integrate the following functions with respect to x:

(a) x^6; (b) $x^{1/3} + 6$; (c) $\sin 2x$; (d) e^{2x}; (e) $1/(ax + b)$; (f) $\sec^2 4x$.

(v)   The contour of an open surface subtends a solid angle of a steradians at a point P. What area does a cone of lines, drawn from the contour to P, intercept on a sphere of radius r drawn about P as centre?

(vi) Determine the turning points of the following functions and state whether they are maximum or minimum values:

(a) $x^2 + 4x$; (b) $(x^3 - 4)/x^2)$; (c) $(x - 3)(2x + 1)$.

47

ALTERNATING QUANTITIES

A quantity whose values are alternately positive then negative over a period of time is said to be an alternating quantity. The instantaneous value of many alternating quantities changes with cyclic regularity and each complete change or cycle takes place in a time called the periodic time (T).

The number of reversals in one second is called the *frequency* of the alternation. The unit of frequency is the Hertz (Hz) and 1 Hz = 1 cycle per second. The graph of the quantity to a base of time is called its *waveform* and two examples are given in the diagram below.

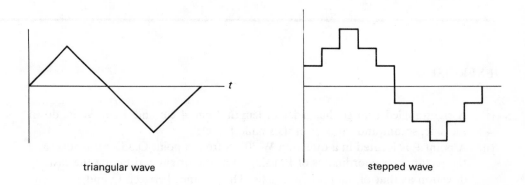

triangular wave   stepped wave

A quantity whose values, though varying with time, are always positive (or always negative) is *not* an alternating quantity but is termed a varying direct quantity.

Alternating voltages and currents have some important advantages over direct voltages and currents and these are summarised below.

(i)   It is easy to change from one level of magnitude to another by means of a device called a transformer. We will study the transformer in some detail in Programme 8.
(ii)   The task of switching off a current is easier if it passes through a zero value periodically.
(iii) The cheapest industrial motor is the three phase induction motor which is an alternating current machine. We will study this type of motor in Programme 9.

What is the relationship between the frequency of the alternation (f) and the periodic time (T)?

48

$$f = 1/T \qquad \text{(Hz)}$$

... since one cycle occupies $T(\text{s})$ there are $1/T$ cycles per second.

A very wide range of frequencies exists in electrical and electronic engineering. For example, for power applications frequencies in the range 12–100 Hz are used; for electrical reproduction of speech and music the range is 30 Hz–18 kHz while for radio, television and radar applications it is 16 kHz–10 GHz.

49

There are a number of different types of 'value' associated with alternating quantities and these are defined as follows:

Instantaneous value: this will in general be different from instant to instant and is denoted by a lower case letter (e.g. v, i, e).

Maximum or *peak value*: this is the highest value reached by the quantity during one complete cycle and is denoted by a subscripted capital letter (e.g. V_m, I_m, E_m).

Mean value: if an alternating quantity has similar positive and negative half cycles, its average value, taken over a complete cycle, is zero. For this reason, the average value over a half cycle is taken. It is denoted by a capital letter with the subscript av (e.g. V_{av}).

Root mean square (r.m.s.) value: this is also often referred to as the effective value. The r.m.s. value of an alternating current is that value which produces the same heating effect as a direct (steady) current of the same value. It is calculated from the square root of the mean squares of the current.

50

SINUSOIDALLY ALTERNATING QUANTITIES

Generators in power stations are designed to produce a voltage whose waveform is very close to a sinusoid. This is because the sine wave is the only alternating waveform which, when differentiated, gives a waveform of the same shape. All other alternating waveforms, when differentiated, yield a succession of higher frequency waveforms, called harmonics, which give rise to extra power losses or higher insulation stresses.

51

A sinusoidal voltage may be represented by:

$$e = E_m \sin(2\pi f t)$$

where e is the instantaneous value of the voltage at time t,

   E_M is the maximum value of the voltage,

and f is the frequency of the alternation.

$2\pi f$ is usually written as ω, and called the *angular frequency* of the alternation. The unit of ω is the radian per second (rad s^{-1}).

E_M and ω are constants and the waveform takes the following shape:

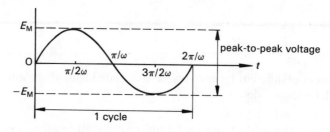

What is (i) the angular frequency, and (ii) the periodic time of a sinusoidal voltage whose frequency is 50 Hz?

52

(i) 314 rad s^{-1} ($2\pi \times 50$); (ii) 0.02 s (1/50)

PHASORIAL REPRESENTATION OF SINUSOIDALLY VARYING QUANTITIES

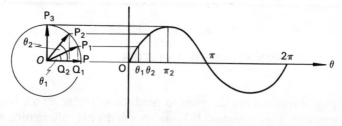

In the diagram, the line OP is considered to be rotating in an anticlockwise direction with a constant angular velocity of ω radians per second. Starting from the horizontal position OP, after θ_1/ω seconds the line will have reached the position OP$_1$; after θ_2/ω seconds it will have reached the position OP$_2$, and after $\pi/2\omega$ seconds it will be in position OP_3.

Now
$$\frac{P_1Q_1}{OP_1} = \frac{P_1Q_1}{OP} = \sin\theta_1.$$

Also
$$\frac{P_2Q_2}{OP_2} = \frac{P_2Q_2}{OP} = \sin\theta_2$$

and
$$\frac{OP_3}{OP} = \sin(\pi/2) = 1.$$

If, therefore, the horizontal projections of the line OP as it moves in a circular path are plotted to a base of θ, the resulting graph will be a sine wave as shown in the diagram. The rotating line is called a *phasor*.

How may the horizontal axis of the diagram be converted into a time scale?

53

> By dividing θ (radians) by the constant angular frequency, ω (radians per second) to give (θ/ω) (s)

PHASOR DIAGRAMS

Suppose that, in a given circuit, a voltage represented by $v = V_m \sin\omega t$ gives rise to a current which is represented by $i = I_m \sin(\omega t + \phi)$. The current is said to have a phase angle ϕ, and you will see that, at a time $t = 0$ (when the voltage is zero) the current has a value $I_M \sin\phi$.

The current and voltage may be represented as phasors and this is done in the following diagrams which are called phasor diagrams.

The information given is:

- the magnitude of the voltage (the length of the line which can be scaled to represent the maximum value or any other constant multiple of it);
- the magnitude of the current (the length of the line to a suitable scale);
- the phase difference between the voltage and the current. In the example above, the current is said to *lead* the voltage or the voltage is said to *lag* the current.

54

Notice that the information given in both diagrams in the previous frame is identical. The lengths of the lines representing I and V and the angle ϕ are the same in both diagrams. The phasors are rotating in an anticlockwise direction at an angular frequency (ω). It is immaterial at what instant we 'stop' the phasors. Only quantities having the same frequency may be represented on the same phasor diagram.

The following diagram represents four voltages. Taking v_1 ($= V_{1m} \sin \omega t$) as the reference, write down expressions for each of the other three.

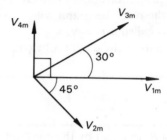

55

$$
\begin{array}{l}
v_2 = V_{2M} \sin(\omega t - \pi/4) \\
v_3 = V_{3M} \sin(\omega t + \pi/6) \\
v_4 = V_{4M} \sin(\omega t + \pi/2)
\end{array}
$$

If we wish to add these four voltages, we can do so using the same techniques as for the addition of vectors. Can you remember these methods?

56

The 'chain' method.
The method of completing the parallelogram of vectors (phasors).
The method of resolving the vectors (phasors) into two components at right angles.
(See Frames 11, 12 and 13.)

57

THE r.m.s. VALUE OF A SINUSOIDAL QUANTITY

Consider a sinusoidally alternating current represented by $i = I_M \sin \theta$ where $\theta = \omega t$. The square of the current is $i^2 = I_m^2 \sin^2 \theta$ and the sum of this over a complete cycle is the integral of this from $\theta = 0$ to 2π(rad).

i.e. $\displaystyle\int_0^{2\pi} I_M^2 \sin^2 \theta \, d\theta$ and the mean of this is $\displaystyle\frac{1}{2\pi} \int_0^{2\pi} I_M^2 \sin^2 \theta \, d\theta$.

The r.m.s. value is obtained by taking the square root of this:

$$\sqrt{\left(\frac{1}{2\pi} \int_0^{2\pi} I_M^2 \sin^2 \theta \, d\theta \right)}$$

In order to integrate $\sin^2 \theta$ we make use of the identity $\sin^2 \theta = (1 - \cos 2\theta)/2$. Using this identity evaluate $\int_0^{2\pi} \sin^2 \theta \, d\theta$.

58

$$\boxed{\pi}$$

Here is the working:

$$\int_0^{2\pi} \sin^2 \theta \, d\theta = \int_0^{2\pi} ((1 - \cos 2\theta)/2) \, d\theta = \frac{1}{2}\left[\theta - \frac{\sin 2\theta}{2} \right]_0^{2\pi} = \frac{2\pi}{2} - 0 = \pi$$

It follows that $\displaystyle\sqrt{\left(\frac{1}{2\pi} \int_0^{2\pi} I_M^2 \sin^2 \theta \, d\theta \right)} = \sqrt{\left(\frac{1}{2\pi} I_M^2 \times \pi \right)}$

$$= I_M/\sqrt{2}$$

r.m.s. values are represented by capital letters so that $\underline{I = I_M/\sqrt{2}}$.

59

THE AVERAGE VALUE OF A SINUSOIDAL QUANTITY

As stated in Frame 49, this is the average value taken over a half cycle so the average value (I_{av}) of a sinusoidally alternating current represented by $i = I_M \sin \theta$ (where $\theta = \omega t$) is given by

$$I_{av} = \frac{1}{\pi} \int_0^{\pi} I_M \sin \theta \, d\theta$$

What do you make it?

60

$$\boxed{2I_M/\pi}$$

Here is the working:

$$\frac{1}{\pi} \int_0^{\pi} I_M \sin \theta \, d\theta = \frac{I_M}{\pi} [-\cos \theta]_0^{\pi} = \frac{I_M}{\pi} [(-\cos \pi) - (-\cos 0)]$$

$$= \frac{I_M}{\pi} [1 - (-1)]$$

THE FORM FACTOR OF A SINE WAVE

The form factor of a waveform is defined to be the r.m.s. value divided by the average value. A square wave has a form factor of 1 and for a triangular wave it is 1.15. What is the form factor of a sinewave?

61

$$\frac{\text{r.m.s. value}}{\text{average value}} = \frac{\text{maximum value}/\sqrt{2}}{\text{maximum value} \times (2/\pi)} = 1.11$$

62

Which of the following statements are true?

(a) An alternating quantity is one whose value changes regularly with time.
(b) The unit of the frequency of alternation is the hertz (Hz).
(c) The hertz is equivalent to the radian per second.
(d) Frequency (f) is the reciprocal of periodic time (T).
(e) 50 sin 314t represents an alternating quantity which has a maximum value of 50 units and a frequency of 314 Hz.
(f) A phasor diagram can only be used to represent sinusoidally alternating quantities of the same frequency.
(g) A phasor diagram can only be used to represent maximum values.
(h) The r.m.s. value of a sinusoidally alternating quantity is $(1/\sqrt{2})$ times its maximum value.
(i) The average value of a sinusoidally alternating quantity is zero.
(j) The form factor of a sine wave is 1.11.

63

$$\boxed{\text{True: b, d, f, h, j}}$$

(a) It must take positive and negative values cyclicly.   (Frame 47)
(c) One hertz is one cycle per second.   (Frame 47)
(e) The frequency is $314/2\pi = 50$ Hz.   (Frame 51)
(g) They can represent r.m.s. values as well.   (Frame 53)
(i) The average value is $2I_M/\pi$.   (Frame 59)

64

SUMMARY OF FRAMES 47–63

Quantity	Symbol	Unit	Abbreviation
Frequency	f	hertz	Hz
Periodic time	$T\,(=1/f)$	second	s
Angular frequency	$\omega\,(=2\pi f)$	radians per second	rad s^{-1}

For a sinusoidal current, $I = I_M/\sqrt{2}$; $I_{av} = 2I_M/\pi$; form factor $= 1.11$.
Sinusoidal quantities can be represented by phasors.

65

EXERCISES

(i) An alternating voltage has the following values over a complete cycle:

from $t = 0$ to π/ω, $v =$ 100 V
from $t = \pi/\omega$ to $2\pi/\omega$, $v = -100$ V

Determine (a) the r.m.s. value of the voltage,
(b) the form factor of the waveform.

(ii) A sinusoidally alternating current is represented by

$$i = 125 \sin(628t + 30°) \text{ A}$$

Determine (a) its r.m.s. value,
(b) its average value,
(c) its phase angle,
(d) its value when $t = 0.015$ s,
(e) the periodic time of the waveform,
(f) the frequency of the alternation.

(iii) A sinusoidally alternating voltage is represented as follows:

$$v = V_M \sin \omega t \text{ (V)}$$

Give (in similar form) expressions for the following sinusoidal currents:
(a) one of maximum value 5 A which leads the voltage by 35° and is of the same frequency as the voltage,
(b) one whose r.m.s. value is 5 A and for which the frequency is three times that of the voltage.
Explain why a phase angle cannot be given in this case.

(iv) Two sinusoidally alternating currents have a phase difference of 60°, the second leading the first. Their respective maximum values are 10 A and 15 A. Give trigonometrical expressions for the two currents, taking the smaller as the reference, and determine the time interval between corresponding zero current values.

66

SHORT EXERCISES ON PROGRAMME 1

The numbers in brackets after each question refer to the frame(s) where the answer (or clues to the answer) may be found.

1. What is the SI unit of mass? (3)
2. What are the dimensions of velocity? (2)
3. What power of ten is represented by 'Giga'? (7)

4. Distinguish between scalar and vector quantities. (9)
5. State the Principle of Superposition as applied to forces. (14)
6. What is the reciprocal of the cosine of an angle? (21)
7. Give an expression in complex form (j-notation) for a general point in the
   third quadrant. (24)
8. Give an expression for the curved surface area of a cylinder of radius r and
   length l. (25)
9. Write down an expression for the volume of a sphere of radius a. (25)
10. What is the solid angle subtended by an open surface at a point in the
    plane of its contour. (28)
11. If $y = u/v$ where u and v are both functions of x, describe how you would
    proceed to determine dy/dx. (32)
12. What is the integral of $1/x$ with respect to x? (37)
13. Give a mathematical expression for 'the line integral of $f(x, y)$ over a
    path L'. (38)
14. Define an alternating quantity. (47)
15. What is meant by the 'waveform' of an alternating quantity? (47)
16. Give the relationship between frequency (f) and periodic time (T). (48)
17. What is a 'phasor'? (52)
18. What is the meaning of 'phase difference'? (53)
19. State the relationship between the r.m.s. value and the maximum value
    of an alternating quantity. (58)
20. Define the form factor of a waveform and give its value in the case of a
    sine wave. (60,61)

67

ANSWERS TO EXERCISES

Frame 19

(i)   (a) $[LMT^{-2}]$; (b) $[L^2MT^{-2}]$; (c) $[L^2MT^{-3}]$; (d) $[L^2MT^{-3}A^{-1}]$.
(ii) (a) 35 mm;   (b) 6 200 kW;   (c) 0.108 mJ;
      (d) 0.5×10^6 pF; (e) 3.3×10^6 mV; (f) 300×10^3 μA.
(iii) 38.72 N acting in a direction W 18.8° N.

Frame 46

(i)   0.6; 1.25; 0.75. (ii) $(5 \angle 150°)$; $(-4.33, 2.5)$; $(-4.33 + j2.5)$.
(iii) (a) $5x^4$; (b) $-4/x^5$; (c) $3(2x \cos 2x + \sin 2x)$; (d) 0;
      (e) $e^x(\sin x - \cos x)/\sin^2 x$; (f) $1 + \ln x$.
(iv) (a) $(1/7)x^7 + C$; (b) $(3/4)x^{4/3} + 6x + C$; (c) $(1/2)(-\cos 2x) + C$;
      (d) $(1/2)e^{2x} + C$; (e) $(1/a)\ln(ax + b) + C$; (f) $(1/4) \tan 4x + C$.
(v) a/r^2.
(vi) (a) -4, minimum; (b) -3, maximum; (c) $-49/8$, minimum.
The answers to the test exercise in Frame 65 are given on the next page.

Frame 65

(i)   (a) 100 V; (b) 1.
(ii) (a) 88.4 A; (b) 79.58 A; (c) 30°; (d) 62.81 A; (f) 0.01 s; (g) 100 Hz.
(iii) (a) $i = 5 \sin(\omega t + 35°)$ A; (b) $i = 7.07 \sin 3\omega t$ (A). Since the phasors are rotating at different speeds, the 'phase angle' will be constantly changing.
(iv) $10 \sin 314t$; $15(\sin 314t + 60°)$; 3.33 ms.

Programme 2

ELECTROSTATIC FIELDS 1

1

INTRODUCTION

Electrostatic field theory is concerned with electric charges at rest. In this programme we begin the study of electrostatics by becoming familiar with the meaning of the following terms and concepts:

- electric charge
- electric field strength
- electric field
- electric lines of force (also known as electric flux or streamlines)
- electric flux density
- absolute and relative permittivity
- electric potential and potential difference
- electric potential gradient
- equipotential surfaces
- electric dipoles
- the moment of electric dipoles.

We shall learn how to:

- use Coulomb's law to calculate the force between point charges
- decide whether the forces between charges are attractive or repulsive
- use Gauss's theorem to relate electric flux to electric charge
- apply the Principle of Superposition to find the resultant effect when a number of forces or charges are acting simultaneously.

When you have studied this programme you should be able to:

- distinguish between vector and scalar quantities commonly used in electrostatic field theory
- calculate the magnitude and direction of the force between point charges
- calculate the magnitude and direction of the force on a charge given its strength and the strength of the electric field in which it finds itself
- calculate the magnitude and direction of the electric field strength at a point in an electric field resulting from point charges, charged spheres, line charges, charged cylinders and electric dipoles
- calculate the electric potential at a point in an electric field due to any of the systems of charges mentioned above
- calculate the electric potential difference between two points in an electric field due to any of these systems of charges
- sketch the field patterns and potential maps due to simple systems of charged bodies.

2

All things, whether solid, liquid or gaseous, animate or inanimate, are made up of a number of atoms which themselves consist of a number of particles. Physicists have discovered many different subatomic particles (more than thirty) but the most important of these so far as we are concerned are the electron, the proton and the neutron.

An electron is a particle whose mass is 9.11×10^{-31} kg and which carries negative electric charge. All electrons are identical so that those which are part of a pint of water are identical to those which are part of the glass holding the water. A proton is a particle carrying positive electric charge whose mass is 1.6×10^{-27} kg. The amount of positive electricity in a proton is exactly the same as the amount of negative electricity of an electron. A neutron is a particle whose mass is the same as that of the proton and which has no electricity. All atoms (with the exception of the hydrogen atom which has one electron and one proton but no neutrons) contain all three particles, the protons and the neutrons together forming the nucleus of the atom.

How many times greater than the electron mass is that of the proton?

3

1789 times (divide 1.63×10^{-27} by 9.11×10^{-31})

ELECTRIC CHARGE

All atoms are normally neutral electrically since the total number of protons equals the total number of electrons. If some of the electrons are removed from the atoms of a thing (let's call it a body), that body will have a deficiency of negative electricity. However, electrons cannot be destroyed so it follows that some other body must have a surplus of electrons.

An atom which has lost electrons is called a positive ion; one which has gained electrons is called a negative ion. A body which has a deficiency of electrons is said to be positively electrified or charged, while one having a surplus of electrons is said to be negatively charged. The total deficiency or surplus of electrons in a body is called the charge and the SI unit of charge is the Coulomb (C). The smallest amount of charge is the electricity of an electron which is 1.6×10^{-19} C.

How many electrons are required to form 1 C of electricity?

4

$$\boxed{6.25 \times 10^{18}}$$

This is obtained by dividing 1 C by the charge on one electron (1.6×10^{-19} C).

Coulomb showed that the forces between point charges obey the inverse square law and, applied to charges, the law is known as *Coulomb's law*. It may be stated mathematically as $F = \dfrac{Q_1 Q_2}{4\pi\varepsilon d^2}$ Newton(N) where F is the mutual force between the charges and is a vector quantity. Q_1 and Q_2 are the magnitudes of the charges which are separated by a distance d, and ε is a constant which depends on the medium surrounding them. We will discuss the constant ε, which is called permittivity, in Frame 20. The force acts along the line joining the points and is one of attraction if one of the charges is positive and the other is negative. However if both charges are negative or both are positive the force is one of repulsion.

What is the sign of F if the force is attractive?

5

$$\boxed{\text{Negative}}$$

Clearly, charges cannot exist at a 'point' though an electron approximates closely to a point charge. The problem is that if the charges are not at points it is difficult to define the distance between them. What we can say, though, is that provided the distance between the charges is large compared with the radii of the charges, then the distance between them can be taken to be the distance between their centres.

The constant ε has a value of 8.854×10^{-12} SI units in a vacuum, when it is written ε_0. In air, at normal temperature and pressure, its value is only 0.06% less and engineers usually use the same value for air as for a vacuum. It follows that in a vacuum, $1/4\pi\varepsilon = 8.988 \times 10^9$ SI units.

Use Coulomb's law to calculate the force between a proton and an electron separated by a distance of 1 micron (10^{-6} m) in air.

6

$$\boxed{2.3 \times 10^{-16} \text{ N (attractive)}}$$

$$\ldots F = \frac{8.988 \times 10^9 \times 1.6 \times 10^{-19} \times (-1.6 \times 10^{-19})}{(10^{-6})^2} \text{ N} = \underline{-2.3 \times 10^{-16} \text{ N}}$$

As we saw in Frame 5, the negative sign indicates that the force between the charges is one of attraction.

This answer assumes that there are no other charges in the vicinity and that gravitational forces are negligible. Is this second assumption reasonable? Well, yes it is. In this particular example the gravitational force between the particles is of the order of 10^{40} times smaller than the electrostatic force.

What are the dimensions of the constant ε which appears in Coulomb's law?

7

$$\boxed{[M^{-1}L^{-3}T^4A^2] \text{ (see Programme 1, Frame 6)}}$$

So we can now calculate the force between two charges in a vacuum or air. What if there are more than two charges? How can we calculate the force on any one of them? Well, we can call upon the Principle of Superposition. This tells us that the force on any one charge is the vector sum of the forces which would be exerted on it by each of the others if they acted alone.

In the diagram F_{12} represents the force on charge Q_1 due to charge Q_2; F_{13} represents the force on Q_1 due to the charge Q_3; F_R represents the resultant force on charge Q_1 and is the vector sum of F_{12} and F_{13}.

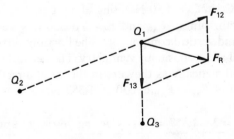

What can you deduce about the signs of the charges Q_1 and Q_2 from the diagram?

8

They are the same because the force on Q_1 is repulsive

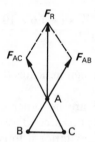

Let us now calculate the force on a charge of 2 μC placed at the apex A of an equilateral triangle of side 10 cm. Charges of 1 μC are placed at the other two vertices B and C. Using the Principle of Superposition, we first calculate the force on the charge at A due to the charge at B. Call this F_{AB}. The charge at C is ignored while we do this.

$$F_{AB} = \frac{Q_A Q_B}{4\pi\varepsilon_0 (AB)^2} = \frac{8.988 \times 10^9 \times 2 \times 10^{-6} \times 1 \times 10^{-6}}{(0.1)^2} \text{ N} = 1.80 \text{ N along } \overrightarrow{BA}.$$

Now you complete the problem.

9

$F_R = 3.11$ N directed vertically upwards

You should have found that, ignoring the charge at B, the force on the charge at A due to the charge at C is $F_{AC} = 1.80$ N acting along $\overrightarrow{CA}$. The resultant force on the charge at A is the vector sum of F_{AB} and F_{AC}. To find this we resolve F_{AB} and F_{AC} into their horizontal and vertical components. The horizontal components will cancel because they are equal and opposite by symmetry. The vertical components are equal and can be added arithmetically as they are in the same direction.

Vertical component of $F_{AB} = F_{AB} \cos 30° = 1.557$ N.

$$\therefore F_R = 2 \times 1.557 = 3.11 \text{ N acting vertically upwards}$$

What would be the force experienced by a charge of 1 C placed at A?

$$\boxed{1.557 \times 10^6 \text{ N}}$$

This is because the force experienced by a charge of 1 C is $1/(2 \times 10^{-6})$ times as great as that experienced by a charge of 2×10^{-6} C.

ELECTRIC FIELD STRENGTH

This leads us to an alternative method of solving problems of this kind. We introduce a vector quantity called Electric Field Strength the symbol for which is E. The electric field strength at a point is defined as the force which would be experienced by a unit positive charge (i.e. a charge of $+1$ C) placed at the point.

In the previous problem, the electric field strength at A due to the charge at B is given by $E_{AB} = \dfrac{Q_B}{4\pi\varepsilon_0(AB)^2} = \dfrac{8.988 \times 10^9 \times 1 \times 10^{-6}}{0.1^2} = 8.988 \times 10^5$ SI units

(acting along $\overrightarrow{BA}$).

Similarly $E_{AC} = \dfrac{Q_C}{4\pi\varepsilon_0(AC)^2} = 8.988 \times 10^5$ SI units along $\overrightarrow{CA}$.

The total electric field strength at A due to both the charges at B acting together is the vector sum of E_{AB} and E_{AC} by the Principle of Superposition. Call this E_R and remember that it must have magnitude and direction for complete specification.

Resolving E_{AB} and E_{AC} into their horizontal and vertical components as before, we find that the horizontal components cancel and the vertical components add. The resultant is thus $2(E_{AB} \cos 30°) = 1.557 \times 10^6$ N and acts vertically upwards as shown in the diagram in Frame 8.

Each unit of charge placed at A will therefore experience a force of 1.557×10^6 N. The force experienced by the charge of 2×10^{-6} C is therefore
$1.557 \times 10^6 \times 2 \times 10^{-6} = \underline{3.11 \text{ N}}$ as before.

Can you think of a general equation for the force on a charge of Q Coulomb at a point where the electric field strength is E SI units?

From this equation you can determine the SI unit of electric field strength.

> $F = QE$. It follows that the unit of E is the unit of
> of F divided by the unit of Q, i.e. the Newton per coulomb (N C^{-1}).

Example

Calculate the electric field strength at a point P midway between two charges Q_A (10 μC) and Q_B (-10 μC) separated by a distance of 1 m in air.

Solution

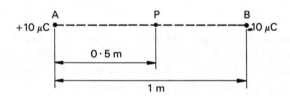

The electric field strength at P due to Q_A at A is given by

$$E_{PA} = \frac{Q_A}{4\pi\varepsilon_0(AP)^2} = \frac{8.988 \times 10^9 \times 10 \times 10^{-6}}{(0.5)^2} \text{ N C}^{-1} \text{ along } \overrightarrow{AP}$$

$$= 359.5 \times 10^3 \text{ N C}^{-1} \text{ along } \overrightarrow{AP}$$

Similarly, the electric field strength at P due to Q_B at B is E_{PB} and this has the value 359.5×10^3 N C^{-1} along $\overrightarrow{PB}$.

The resultant field strength at P is obtained using the Principle of Superposition and is the vector sum of E_{PA} and E_{PB}. Since these vectors are co-linear we simply add them algebraically giving

$$E_R = E_{PA} + E_{PB} = (359.5 + 359.5) \times 10^3 \text{ N C}^{-1} \text{ along } \overrightarrow{AP}$$

$$= 719 \times 10^3 \text{ N C}^{-1} \text{ along } \overrightarrow{AP}$$

What would be the force experienced by charges of (a) 2 μC, (b) -3 μC, placed at P?

12

> (a) 1.438 N directed along $\overrightarrow{AP}$
> (b) 2.157 N directed along $\overrightarrow{PA}$

Each positive unit of charge experiences a force of 719×10^3 N urging it along $\overrightarrow{AP}$. Each unit of negative charge will therefore experience a force in the opposite direction, that is, along $\overrightarrow{PA}$.

ELECTRIC FIELDS AND LINES OF FORCE

The region in which forces are experienced due to the presence of electric charges is called an electric field. At every point within this region the electric field strength will have both a magnitude and a direction, and to represent the electric field diagrammatically we draw lines of force which are also known as electric flux lines or streamlines. These lines which map out an electric field must always bear arrow-heads to indicate their direction.

PROPERTIES OF LINES OF FORCE

A line of force is defined as a line whose direction at every point is the direction of the resultant *E*-vector at that point. Since the resultant electric field strength at a point cannot have two directions, it follows that lines of force cannot intersect.

Lines of force are close together where the electric field is strong and widely separated where the field is weak.

Parallel lines of force in the same direction repel one another while parallel lines of force in opposite directions attract one another.

A positive charge placed at a point in an electric field will experience a force whose direction is the direction of the line of force at that point.

A line of force leaves a positively charged body and enters a negatively charged body.

These properties are illustrated in the following two frames.

13

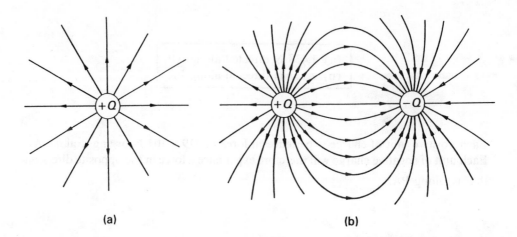

(a)   (b)

Diagram (a) shows lines of force due to an isolated charged sphere. They *leave* the sphere which is positively charged and the further from the sphere they are, the further apart they become as they diverge and the field becomes weaker. They proceed to infinity since the sphere is assumed to be isolated. Diagram (b) shows the field pattern due to two oppositely charged spheres. The lines of force *leave* the positively charged sphere and *enter* the negatively charged sphere.

These diagrams are necessarily two-dimensional although of course in reality the lines of force will emerge from and enter the spheres in three dimensions. The following frame shows the result of replacing the positive charge in each case with a negative one.

14

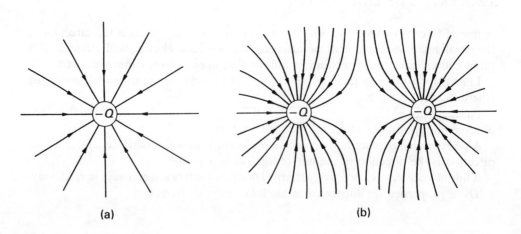

(a)   (b)

15

Now decide whether the following statements are true or false:

(a) All electrons are identical regardless of the atom to which they belong.
(b) The mass of a proton is negligible compared with that of an electron.
(c) An atom which has lost an electron is called a negative ion.
(d) The nucleus of an atom is made up of a proton and one or more electrons.
(e) The total deficiency or excess of electrons in a body is called its charge.
(f) The smallest amount of known charge is that on an electron.
(g) An atom approximates to a point charge.
(h) Coulomb's law strictly applies only to point charges.
(i) The force between point charges is inversely proportional to the square of the distance between them.
(j) The SI unit of charge is the Coulomb.
(k) The unit of electric field strength is the Newton–Coulomb.
(l) Coulomb's law only applies to charges in a vacuum.
(m) A line of force gives the direction of the E-vector at every point in an electrostatic field.
(n) Like charges repel; unlike charges attract.
(o) Lines of force are closer together where the field is weaker.
(p) The Principle of Superposition states that the resultant effect of a number of forces acting together is the same as if each force were considered to be acting alone in turn and the individual effects were added vectorially.
(q) The force on a charge of Q coulomb situated in an electric field of strength E is given by $F = Q/E$.
(r) An electric field is a region in which a charge experiences a force.

16

Tue: a, e, f, h, i, j, m, n, p, r

(b) A proton is much heavier than an electron.   (Frame 2)
(c) Such an atom is called a positive ion.   (Frame 3)
(d) The nucleus consists of protons and neutrons.   (Frame 2)
(g) An electron approximates to a point charge.   (Frame 5)
(k) The unit is the Newton per Coulomb.   (Frame 11)
(l) It applies in other media, too.   (Frame 4)
(o) They are closer together where the field is stronger.   (Frame 12)
(q) $F = QE$.   (Frame 11)

17

SUMMARY OF FRAMES 1–16

Quantity	Symbol	Unit	Abbreviation
Charge	Q	Coulomb	C
Force	F	Newton	N
Electric field strength	E	Newton per Coulomb	$N\,C^{-1}$

LAWS AND EQUATIONS

$$F = \frac{Q_1 Q_2}{4\pi\varepsilon_0 d^2} \text{ (Coulomb's Law)}$$

$F = QE$ (force on a charge Q in a field of strength E)

The Principle of Superposition can be applied to the effects of a number of charges acting together and to a number of forces acting together.

18

EXERCISES

(i) How much negative electricity is present in 10 hydrogen atoms?

(ii) How many protons are required to make up $+0.5$ C of electricity?

(iii) A charge of 10 μC is situated at the origin. Calculate the electric field strength: (a) 2 m along the $(+X)$-axis, (b) 3 m along the $(-Y)$-axis

(iv) Charges of -20 nC, 10 nC, -30 nC and 50 nC are placed on the corners A, B, C and D respectively of a rectangle for which $AB = CD = 10$ cm and $BC = DA = 5$ cm. Determine the force experienced by a charge of -2 nC placed at the centre of side AB.

(v) Charges of 20 nC and -20 nC are placed 0.1 m apart in air at points A and B respectively. Determine the electric field strength at a point in the line of the charges and 0.04 m outside A (0.14 m from B).

(vi) Sketch the field pattern due to the charges of question (v).

19

ELECTRIC FLUX DENSITY

We have seen that a charged body has a region around it in which other charged bodies experience a force. We called this region a field and to picture its form we imagined it to be filled with lines of force or flux whose direction at every point is the direction of the *E*-vector at that point. The closer together the flux lines are, the stronger is the field.

It is found to be convenient to use a new vector, called the electric flux density, whose symbol is *D* and which is intimately connected with the electric field strength.

In fact it is defined as $$D = \frac{E}{\varepsilon_0} \qquad D = \varepsilon_0 E$$

This means that *D* is a vector whose direction is the same as that of *E* and which is a constant multiple of it.

You are going to have to use this 'constant multiple' many times during this (and the following) programme so it would be as well to try to remember its numerical value. Can you remember it?

20

$$\boxed{8.854 \times 10^{-12}}$$

PERMITTIVITY

As was mentioned in Frame 4 in connection with Coulomb's law, the constant ε is dependent upon the medium of the field and is therefore a characteristic of it. It is called the absolute permittivity of the medium of the field or the electric constant. For a vacuum it is called the absolute permittivity of free space or the primary electric constant. In this case it is given the special symbol ε_0. Incidentally, ε is the Greek letter epsilon.

Derive a unit for ε_0.

21

$$\boxed{\text{The (Coulomb)}^2 \text{ per Newton (metre)}^2 \ (\text{C}^2 \ \text{N}^{-1} \ \text{m}^{-2})}$$

Since $F = Q_1 Q_2 / 4\pi\varepsilon_0 d^2$, then $\varepsilon_0 = Q_1 Q_2 / 4\pi F d^2$ and so the unit of ε_0 is (the unit of $Q_1 \times$ the unit of Q_2) divided by (the unit of $F \times$ the unit of d^2).

ELECTRIC FLUX

The flux lines in a field will in general be changing in direction and density from place to place. They will also form a three-dimensional picture.

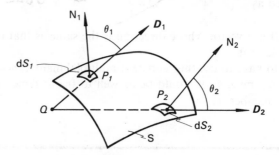

The diagram shows a curved surface S near a charge Q. The lines of force due to Q will cross the surface in different directions and two of them are shown. They cross S at points P_1 and P_2 which are on small areas (dS_1 and dS_2 respectively) of S. If dS_1 and dS_2 are small enough they can be considered to be flat. $P_1 N_1$ is normal (i.e. at right angles to) dS_1 and $P_2 N_2$ is normal to dS_2. There will be an infinite number of small areas like dS_1 and dS_2 which go to make up S.

What is the normal component of the flux density vector D_1 at P_1?

22

$$\boxed{D_1 \cos\theta_1}$$

Similarly the normal component of the flux density vector D_2 at P_2 is given by $D_2 \cos\theta_2$. All of the other surfaces like dS_1 which go to make up the whole surface S will each have a component of electric flux density at right angles to a point such as P_1 on it.

$D_1 \cos \theta_1 \, dS_1$ is called 'the flux across the surface dS_1' (flux density times area equals flux). Similarly $D_2 \cos \theta_2 \, dS_2$ is the flux across the surface dS_2. The symbol for electric flux is ψ (the Greek letter psi). For every area like dS_1 there will be a corresponding $D \cos \theta \, dS$, and the flux across the whole surface S is the sum of all of these. This is written as a surface integral as follows:

$$\int_S D \cos \theta \, dS \qquad \text{or} \qquad \int\int_S D \cos \theta \, dS$$

Just think of it as a short-hand way of saying 'add up all the products of $D \cos \theta$ times dS over the whole surface S'.

In this programme the surface integrals which you will have to evaluate will be quite straightforward.

Example

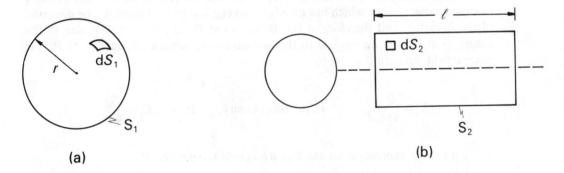

(a)   (b)

The spherical surface S_1 shown in the diagram (a) and the curved cylindrical surface S_2 shown in diagram (b) may each be thought of as being made up of an infinite number of small surfaces like dS_1 and dS_2 respectively.

Evaluate the surface integrals $\int_{S_1} dS_1$ and $\int_{S_2} dS_2$.

Solution

(a) $\int_{S_1} dS_1$ means 'add up all the areas like dS_1 which make up the surface S_1'. This is simply the surface area of a sphere of radius r and amounts to $4\pi r^2$.

(b) $\int_{S_2} dS_2$ is similarly the curved surface area of a cylinder of radius x and length l. This is $2\pi x l$.

23

GAUSS'S THEOREM

This is one of the most important theorems in electromagnetism and we will be using it time and time again throughout this and the following programmes. It is derived here but the important thing is that you are able to *apply* the theorem as stated in Frame 25.

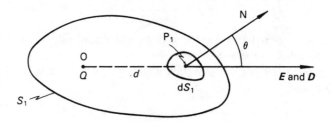

The closed surface shown is made up of an infinite number of small areas like dS_1 (a closed surface is one which has no edge – an egg is a good example). The direction of the E-vector (and therefore of the D-vector) at P_1, a point on dS_1, due to the charge Q at O is at an angle θ to the normal to the surface dS_1 at P_1. At P_1 the electric field strength is given by

$$E_1 = \frac{Q \times 1}{4\pi\varepsilon_0 d^2} \quad \text{and the flux density} \quad D_1 = \varepsilon_0 E_1 = \frac{Q}{4\pi d^2}.$$

Write down an expression for the flux $d\psi$ across the surface dS_1.

24

$$\boxed{d\psi = D_1 \cos \theta \, dS_1}$$

... and since

$$D_1 = \frac{Q}{4\pi d^2}$$

then

$$d\psi = \frac{Q \cos \theta \, dS_1}{4\pi d^2} \tag{i}$$

The next diagram concentrates on dS_1. In the diagram, O, Q, d, P_1, dS_1, N, θ, E and D are all as in the previous diagram.

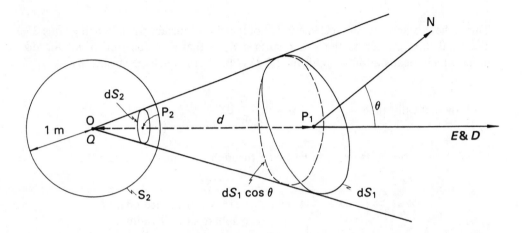

A sphere S_2 of radius 1 m is drawn with centre O so that it encloses Q. If O is joined to every point on the perimeter of dS_1 by means of straight lines, a cone is formed with its apex at O. The area $dS_1 \cos \theta$ is a section of the cone through P_1 and at right angles of OP_1. dS_2 is the area intercepted on the sphere S_2 by the cone of lines.

We are going to show that the flux crossing any closed surface is the same as that crossing the sphere of unit radius enclosing the same charge and is in fact equal to the charge enclosed.

Now since the areas of similar figures are proportional to the square of their linear dimensions we may write

$$\frac{dS_2}{dS_1 \cos \theta} = \frac{(OP_2)^2}{(OP_1)^2} = \frac{1^2}{d^2}$$

(ii)

$$dS_2 = \frac{dS_1 \cos \theta}{d^2}$$

Using (i) and (ii), write down an expression for the flux ($d\psi$) across dS_1 in terms of dS_2.

25

$$\boxed{d\psi = Q\, dS_2/4\pi}$$

This is the flux across a small element dS_1 of the closed surface S_1 shown in Frame 23. The total flux (ψ) across the whole surface S_1 is therefore the sum of all the $d\psi$ terms which means adding up all the $(Q\, dS_2/4\pi)$ terms over the surface S_2.

Mathematically then $$\psi = \int_{S_2} \frac{Q}{4\pi}\, dS$$

Q and 4π are constants and can be taken through the integral sign so that

$$\psi = \frac{Q}{4\pi} \int_{S_2} dS = \frac{Q}{4\pi} 4\pi 1^2 \qquad \begin{array}{l}(4\pi 1^2 \text{ is the surface area of}\\ \text{a sphere of unit radius)}\end{array}$$

$$\therefore\ \underline{\psi = Q}$$

If there are more than one charge enclosed by the surface S_1 we use the Principle of Superposition to show that the flux ψ crossing the surface is the sum of the charges enclosed.

i.e. $$\psi = \Sigma Q$$

This is Gauss's theorem and is stated as follows:

'The net electric flux across (*leaving*) any closed surface is equal to the charge enclosed by that surface'.

What is the flux across the surface S in each of the following diagrams?

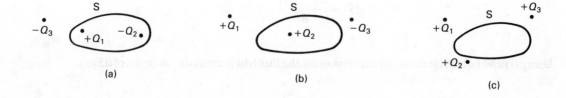

(a)　　　　　(b)　　　　　(c)

26

> (a) $Q_1 - Q_2$;   (b) Q_2;   (c) 0

The important points to note are:

(i)   the theorem only applies to *closed* surfaces;
(ii)   any charges outside the surface do not count because any flux *leaving* the surface at some point due to them must have *entered* the surface (i.e. crossed it in the opposite direction) somewhere else, the *net* flux *leaving* the surface being nil.

UNIT OF ELECTRIC FLUX

Since Q units of charge produce Q units of flux across a surface, the same unit is used for both. The SI unit of electric flux is therefore the coulomb.

What, then, is the unit of electric flux density?

27

> The coulomb per square metre ($C\ m^{-2}$)

Example

Use Gauss's theorem to obtain an expression for the electric field strength E_x at a point distant x from the centre of a sphere of radius r ($r < x$) if the sphere has a uniformly distributed surface charge of Q coulomb and the medium of the field is air.

The steps in the solution are, in reverse order:

- to obtain E_x we can use $E_x = D_x/\varepsilon_0$, therefore we need D_x;
- to obtain D_x, the flux density at x, we can use $D_x = \psi/A$ where ψ is the flux existing at distance x from the centre of the sphere and A is the area across which it exists;
- to obtain ψ we can use Gauss's theorem which relates ψ to the charge Q;
- to use Gauss's theorem we need a 'Gaussian surface' to which we can apply the theorem.

We will now solve the problem carrying out these steps in the appropriate order. First suggest a suitable Gaussian surface.

28

> A spherical surface of radius x, whose centre
> is the centre of the charged sphere

The Gaussian surface chosen will be dictated by the geometry of the problem under consideration. In the present case we need to know about conditions distant x from the centre of a sphere, so a spherical surface of that radius suggests itself.

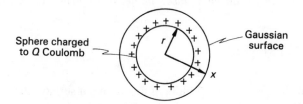

Using Gauss's theorem, determine the flux crossing the Gaussian surface.

29

> $\psi = Q$

The net flux (ψ) across the Gaussian surface at x is equal to the charge enclosed by the surface (Q).

The next step is to obtain the flux density D_x at radius x. Do that now.

30

> $D_x = Q/4\pi x^2$

The electric flux density at distance x is given by $D_x = \psi/4\pi x^2$. But $\psi = Q$ from step 1 so that $D_x = Q/4\pi x^2$.

The final step is to use the relationship between the flux density D_x and the electric field strength E_x to obtain the required expression for E_x.

What do you make it?

31

$$E_x = \frac{Q}{4\pi x^2 \varepsilon_0}$$

... simply using $E = D/\varepsilon_0$.

The direction of the E-vector is radially outwards from points on the surface of the sphere.

What would be the answer for $x < r$?

32

$$E_x = 0$$

A Gaussian surface drawn anywhere inside the charged spherical surface would enclose no charge, so that the net flux crossing it would be zero. It follows that $D_x = 0$ and $E_x = 0$.

Now decide which of the following statements are true:

(a) Permittivity is the ratio of flux density to field strength.
(b) D has the same direction as E and is proportional to it.
(c) The unit of flux density is the Coulomb per metre squared.
(d) ε_0 has a unit equivalent to the $C^2 \ m^{-2} \ N^{-1}$.
(e) Gauss's theorem states that the net flux across a closed surface is zero.
(f) A closed surface is one having no edge or contour.
(g) Gauss's theorem can only be applied to a single point charge.
(h) Gauss's theorem implies that the electric flux density at a point perpendicularly r metre from the axis of a cylinder of radius a charged to q Coulomb per metre length is given by $q/2\pi r$, provided that $r > a$.
(i) Electric flux is D divided by the area over which it acts.

33

True: a, b, c, d, f, h

(e) It is equal to the charge enclosed by the surface. (Frame 25)
(g) It applies equally if there are many charges. (Frame 25)
(i) Electric flux is D multiplied by the area over which it acts. (Frame 22)

34

SUMMARY OF FRAMES 19–33

Quantity	Symbol	Unit	Abbreviation
Electric flux density	D	Coulomb per square metre	$C\,m^{-2}$
Permittivity	ε	to be determined	
Electric flux	ψ	Coulomb	C

LAWS AND EQUATIONS

Coulomb's law: $F = \dfrac{Q_1 Q_2}{4\pi\varepsilon_0 r^2}$ (N); Gauss's theorem: $\displaystyle\int D \cos\theta \, dS = \Sigma Q$;

$D = \varepsilon_0 E$.

35

EXERCISES

(i) Show that the electric field strength at a point P distant d from a point charge of strength Q in a vacuum is given by $E = Q/4\pi\varepsilon_0 d^2$.
(ii) Calculate the maximum charge per metre length which a straight wire of radius 0.5 cm can have if the electric field strength at its surface must not exceed 3×10^6 N C^{-1}.
(iii) A sphere of radius 5 cm has a uniformly distributed surface charge of 50 μC. Plot a graph of electric field strength E_x to a base of the distance x from the centre of the sphere.

ELECTRIC POTENTIAL

If a unit positive charge is free to move in an electric field it will do so under the influence of the E-vector, and the force that it experiences is given by $F = QE$ which becomes $F = E$ if Q is unity. To move the charge against the force exerted by the field requires an external force which must do work.

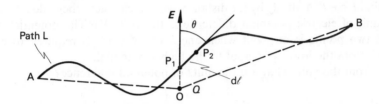

The diagram shows two points A and B situated in an electric field produced by a charge Q at O. The work which must be done by an external force in moving a unit positive charge from A to B is called the increase in electrical potential between A and B.

L is an arbitrary path drawn between A and B, and P_1 and P_2 are two points on this path separated by a very small distance dl. The work done by the external force (call it F) in moving the unit positive charge from P_1 to P_2 is the product of the component of the force in the direction of motion times the distance moved, i.e. $-F \cos \theta \, dl$. The minus sign indicates that the motion is in the direction in which the field itself urges the unit positive charge, so that the work done by the external force is negative.

Since $F = QE$ and $Q = 1$ then the work done is $-E \cos \theta \, dl$ and by definition this is the increase in potential between points P_1 and P_2.

The potential difference between points A and B is found by adding all the products of $E \cos \theta$ with dl for all the small distances like dl which go to make up the path L.

This is usually written as a line integral as follows:

$$\int_L E \cos \theta \, dl$$

The 'L' identifies the path over which the integration takes place.

Using this notation, write down an expression for the potential of point B (call it V_B) with respect to that at point A (call it V_A).

37

$$V_B - V_A = -\int_L E \cos \theta \, dl$$

Since the work done is negative, then the increase in potential is negative so that V_B is lower than V_A. The negative sign indicates that the potential *rises* as the charge is moved against the direction of the field.

$\int_L E \cos \theta \, dl$ is read 'the line integral of E along the path L'. Just think of it as a short-hand way of saying 'for every small distance dl which goes to make up the path L find $E \cos \theta$, multiply by the distance over which it acts, then add all the results'.

The unit of electric potential difference is the volt (V). The potential difference between two points is one volt when one Joule of work is required to move one Coulomb from the lower potential to the higher potential.

Apart from the volt, what are the units of potential difference?

38

The unit of $E \times$ the unit of length (N C^{-1} m)

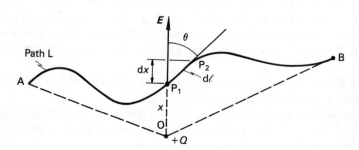

We have that in the diagram

$$V_B - V_A = -\int_L E \cos \theta \, dl = -\int_L \frac{D}{\varepsilon_0} \cos \theta \, dl.$$

Now imagine a spherical Gaussian surface, centre O, radius x, to be drawn through P_1. According to Gauss's theorem the flux, ψ, across this surface is equal to the charge, $+Q$, enclosed by the surface.

Thus $V_B - V_A = -\int_L \dfrac{\psi}{4\pi\varepsilon_0 x^2} \cos\theta\, dl = -\int_L \dfrac{Q}{4\pi\varepsilon_0 x^2} \cos\theta\, dl$ (V)

Taking all the constants through the integral sign and noting that

$\cos\theta\, dl = dx$ we get $V_B - V_A = -\dfrac{Q}{4\pi\varepsilon_0}\int_L \dfrac{dx}{x^2} = -\dfrac{Q}{4\pi\varepsilon_0}\left[-\dfrac{1}{x}\right]_{OA}^{OB}$ (V)

The limits of the integration are the two ends of the path L. At one end (A), $x = OA$ and at the other end (B), $x = OB$.

Thus $$V_B - V_A = \dfrac{Q}{4\pi\varepsilon_0}\left[\dfrac{1}{OB} - \dfrac{1}{OA}\right] \text{ (V)}$$

Can you draw an interesting and very important conclusion from this result? (*Hint*: apart from the source of the field, Q, what does the potential difference between A and B depend upon?)

39

> The potential difference between A and B is independent of the path between the two points. It is therefore a scalar quantity.

Example

Calculate the potential difference between two points A and B which are respectively 10 cm and 5 cm from a point charge of 20×10^{-9} C.

Solution

In this case $Q = 20 \times 10^{-9}$ C, $OA = 10 \times 10^{-2}$ m, $OB = 5 \times 10^2$ m.

$$V_B - V_A = \dfrac{20 \times 10^{-9}}{4\pi \times 8.854 \times 10^{-12}}\left[\dfrac{1}{5 \times 10^{-2}} - \dfrac{1}{10 \times 10^{-2}}\right] = \underline{1797.5 \text{ V}}\ (V_B \text{ higher})$$

What would be the effect if the charge were negative?

40

> V_A would be higher than V_B

The expression we have obtained for the potential difference between two points is for an electric field produced by a single point charge. If there are n charges we use the Principle of Superposition. This means calculating the potential difference between the two points due to each charge considered to be acting alone and then adding the results algebraically.

If the charges causing the field are $Q_1, Q_2, \ldots, Q_n$ and the distances from each one to points A and B are $d_{A1}, d_{A2}, \ldots, d_{An}$ and $d_{B1}, d_{B2}, \ldots, d_{Bn}$ respectively, write down an expression for the potential difference between the two points.

41

$$V_B - V_A = \sum_{m=1}^{n} \frac{Q_m}{4\pi\varepsilon_0}\left[\frac{1}{d_{Bm}} - \frac{1}{d_{Am}}\right] \text{(V)}$$

Example

Calculate the potential difference between point A (3,4) and point B at the origin situated in an electric field produced by a charge Q_1 (of 20 μC) at point (0,4) m and a charge Q_2 (of -10 μC) at point (3,0) m.

Solution

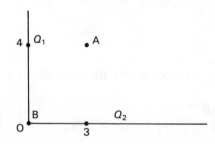

The arrangement is shown in the diagram above.

We use the formula in the box at the top of this frame in which $n = 2$, $Q_1 = 20 \times 10^{-6}$ C, $Q_2 = -10 \times 10^{-6}$ C, $d_{B1} = 4$ m, $d_{B2} = 3$ m, $d_{A1} = 3$ m, $d_{A2} = 4$ m and $\varepsilon_0 = 8.854 \times 10^{-12}$ SI units.

$$V_B - V_A = \frac{20 \times 10^{-6}}{4\pi\varepsilon_0}\left[\frac{1}{4} - \frac{1}{3}\right] + \frac{-10 \times 10^{-6}}{4\pi\varepsilon_0}\left[\frac{1}{3} - \frac{1}{4}\right] \text{ V}$$

$$= \underline{-22.47 \text{ kV}} \text{ (i.e. point A is at the higher potential)}$$

42

ELECTRIC POTENTIAL GRADIENT

The two points A and B in the diagram are separated by a short distance dx:

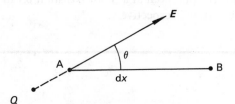

We have that $V_B - V_A = -E \cos \theta \, dx$.

$E \cos \theta \, dx$ is the component of E in the X-direction and is written E_x so that $V_B - V_A = E_x \, dx$ (E_x could also mean 'the field strength at point x'. The context will make it clear which is meant.) If the potential difference between A and B is written as ∂V then $E_x = -\dfrac{\partial V}{\partial x}$ and $\dfrac{\partial V}{\partial x}$ is called the potential gradient in the X-direction. The partial differentiation notation is used because electric fields are essentially three dimensional and the potential may well be varying in the Y- and Z-directions as well and there will be potential gradients $\dfrac{\partial V}{\partial y}$ and $\dfrac{\partial V}{\partial z}$.

Summarising, then, we have

$$E_x = -\frac{\partial V}{\partial x}; \quad E_y = -\frac{\partial V}{\partial y}; \quad E_z = -\frac{\partial V}{\partial y}$$

What is the unit of potential gradient?

43

$$\boxed{\text{The volt per metre } (V\ m^{-1})}$$

THE POTENTIAL AT A POINT DUE TO A NUMBER (n SAY) OF CHARGED PARTICLES

So far we have talked about potential difference. If we wish to state the potential 'at a point' then we need to define a datum. We could take zero potential to be at infinity where the field is zero.

Let us then determine the potential difference between a point B and a point A and let A be at infinity so that it is at zero potential.

$$V_B - V_A = V_B - 0 = \sum_{m=1}^{n} \frac{Q_m}{4\pi\varepsilon_0}\left[\frac{1}{d_{Bm}} - \frac{1}{\infty}\right] \quad \text{and} \quad V_B = \sum_{m=1}^{n} \frac{Q_m}{4\pi\varepsilon_0}\left[\frac{1}{d_{Bm}}\right] (V)$$

Use this to determine the potential at a point B distant 0.1 m and 0.5 m from point charges of 20 nC and -10 nC respectively.

44

$$\boxed{1618\ V}$$

This is calculated as follows:

$$V_B = \frac{1}{4\pi\varepsilon_0}\sum\frac{Q}{d_B} = \frac{10^{-9}}{4\pi\varepsilon_0}\left[\frac{20}{0.1} - \frac{10}{0.5}\right] = 1618\ V$$

Example

Charges of 20 nC and -20 nC are placed 0.1 m apart in air at points A and B respectively. Determine:

(a) the potential at a point in the line of the charges and 0.04 m outside A (i.e. 0.14 m from B);
(b) the potential at point C, the apex of an equilateral triangle having AB as its base;
(c) the charge which must be placed at C in order that the potential at point D, the mid-point of CB, shall be 2 kV.

Solution

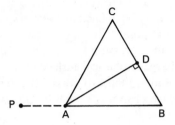

(a) PA = 0.04 m; PB = 0.14 m;
Q_A = 20 nC: Q_B = −20 nC.

$$\therefore\ V_P = \frac{20 \times 10^{-9}}{4\pi\varepsilon_0}\left[\frac{1}{0.04} - \frac{1}{0.14}\right] = \underline{3210\ V}$$

(b) CA = CB = 0.1 m;
Q_A = 20 nC; Q_B = −20 nC.

$$V_C = \frac{20 \times 10^{-9}}{4\pi\varepsilon_0}\left[\frac{1}{0.1} - \frac{1}{0.1}\right] = 0$$

(c) DA = 0.1 cos 30° = 0.0866 m (alternatively DA = $(0.1^2 - 0.05^2)^{1/2}$).
DB = DC = 0.05 m.

$$\therefore\ V_D = \frac{1}{4\pi\varepsilon_0}\left[\frac{20 \times 10^{-9}}{0.0866} + \frac{-20 \times 10^{-9}}{0.05} + \frac{Q_C}{0.05}\right]$$

$$2000 = 8.988 \times 10^9\left[230 \times 10^{-9} - 400 \times 10^{-9} + \frac{Q_C}{0.05}\right]$$

$$Q_C = \left[\frac{2000}{8.988} + 400 - 230\right]0.05 \times 10^{-9}\ C = \underline{19.6\ nC}$$

46

THE ELECTRIC POTENTIAL AT A POINT

There is no absolute value for 'the potential at a point'. It depends upon where we take a datum for zero potential. However, taking zero potential to be at infinity, then 'the potential at a point' may be defined as the work required to bring a unit positive charge from infinity to the point.

We shall now obtain an expression for the potential at a point distant d from the centre of a uniformly charged spherical surface of radius r ($<d$). If we let the charge on the sphere be Q Coulomb, then by Gauss's theorem the flux across the Gaussian surface (a spherical surface of radius x) will be Q Coulomb.

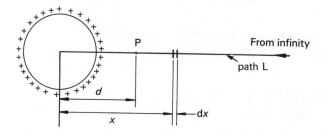

The flux density at x is given by

$$D_x = \frac{\psi}{4\pi x^2} = \frac{Q}{4\pi x^2} \text{ (C m}^{-2})$$

The electric field strength at x,

$$E_x = \frac{D_x}{\varepsilon_0} = \frac{Q}{4\pi x^2 \varepsilon_0} \text{ (N C}^{-1})$$

The work required to move a unit positive charge along an element dx of the path L is $-E \cos\theta \, dx = -E_x \, dx$.

The work required to bring a unit positive charge from infinity to the point P therefore (which by definition is the potential at point P) is given by

$$V_P = -\int_\infty^d E_x \, dx = -\int_\infty^d \frac{Q}{4\pi\varepsilon_0 x^2} \, dx \text{ (V)}$$

$$= \left[\frac{Q}{4\pi\varepsilon_0 x} \right]_\infty^d = \frac{Q}{4\pi\varepsilon_0 d} \text{ (V)}$$

If the spherical surface is uniformly charged to 20×10^{-9} C and its radius is 30 cm, what is the potential at the surface?

Assume that the medium is air.

$$599 \text{ V} \; (20 \times 10^{-9}/(4\pi\varepsilon_0 \times 0.3))$$

THE ELECTRIC POTENTIAL AT A POINT DUE TO AN ELECTRIC DIPOLE

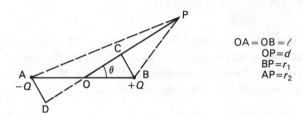

$$OA = OB = l$$
$$OP = d$$
$$BP = r_1$$
$$AP = r_2$$

An electric dipole is a pair of equal and opposite charges separated by a distance which is small compared with the distance at which we wish to measure its effects. The point P in the diagram is specified by the polar co-ordinates $d \angle \theta°$ metre from the centre, O, of the dipole.

Using the expression at the end of the previous frame together with the Principle of Superposition, we have for the potential at point P:

$$V_P = \frac{1}{4\pi\varepsilon_0}\left[\frac{Q}{r_1} - \frac{Q}{r_2}\right] \text{ (V)}$$

We can simplify this by making some approximations and to this end we construct the perpendiculars from A and B on to OP as shown. Then:

$$OC = OD = l \cos\theta$$

Since $d \gg 2l$ then $r_1 \approx d - l \cos\theta$

and $r_2 \approx d + l \cos\theta$.

Substituting in the expression for V_P we obtain:

$$V_P = \frac{Q}{4\pi\varepsilon_0}\left[\frac{1}{(d - l \cos\theta)} - \frac{1}{(d + l \cos\theta)}\right] = \frac{Q}{4\pi\varepsilon_0}\left[\frac{2l \cos\theta}{d^2 - l^2 \cos^2\theta}\right] \text{ (V)}$$

Again, since $d \gg 2l$ then $d^2 \gg l^2 \cos^2\theta$ and $l^2 \cos^2\theta$ may be neglected.

$$\therefore V_P = \frac{Q}{4\pi\varepsilon_0}\left[\frac{2l \cos\theta}{d^2}\right] = \frac{2lQ \cos\theta}{4\pi\varepsilon_0 d^2} = \frac{m \cos\theta}{4\pi\varepsilon_0 d^2} \text{ (V)}$$

where $m = 2lQ$ and is called the *moment* of the electric dipole.

What is the unit of m?

Coulomb metre (C m)

THE ELECTRIC POTENTIAL AT A POINT P PERPENDICULARLY DISTANT d FROM AN INFINITELY LONG STRAIGHT LINE CHARGE OF q COULOMB PER METRE LENGTH

We imagine a 1 m long cylindrical Gaussian surface of radius x to be placed such that the charged line is on its axis. By Gauss's theorem the flux across this surface is equal to q Coulomb. Note that no flux crosses either of the flat ends of the closed Gaussian surface since all the flux lines are radial.

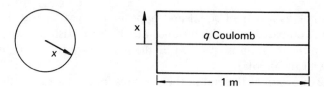

The flux density at radius x is found by dividing the flux across the curved surface of a cylinder of radius x by the area of that curved surface

i.e. $$D_x = \psi/2\pi x = q/2\pi x.$$

The electric field strength at x,

$$E_x = \frac{D_x}{\varepsilon_0} = \frac{q}{2\pi\varepsilon_0 x} \text{ (V m}^{-1})$$

The potential at P $\quad V_P = -\int_z^d E_x \, dx = -\int_z^d \frac{q}{2\pi\varepsilon_0 x} \, dx \text{ (V)}$

where z is the distance to the point where the potential is zero. We cannot use infinity this time because the line itself is infinitely long.

$$V_P = -\frac{q}{2\pi\varepsilon_0}[\ln x]_z^d = -\frac{q}{2\pi\varepsilon_0}(\ln d - \ln z) \text{ (V)}$$

The term $\dfrac{q}{2\pi\varepsilon_0}\ln z$ is the same for all values of d and so the term $-\dfrac{q}{2\pi\varepsilon_0}\ln d$ is taken to be the effective potential at P.

If there are n charged lines in parallel, use the Principle of Superposition to obtain an expression for the effective potential at a point P.

$$V_P = -\frac{1}{2\pi\varepsilon_0} \sum_{a=1}^{n} q_a \ln d_a$$

$q_1, q_2, \ldots, q_n$ are the charges per metre on the n lines and $d_1, d_2, \ldots, d_n$ are their perpendicular distances to the point P.

EQUIPOTENTIAL SURFACES

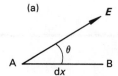

(a)   (b)

We saw in Frame 42 that for diagram (a) above the potential difference between points A and B (call it dV) is given by $dV = -E_x\, dx$ where $E_x = E \cos\theta$. It followed that $E_x = -dV/dx$.

If $\theta = 90°$ as in diagram (b) above, then $E_x = 0$ and so the potential gradient along the line AB is zero. In other words, all points on the line AB are at the same potential and AB is said to be an equipotential line. However, because fields are essentially three dimensional there will be, in general, equipotential surfaces rather than equipotential lines.

What can you say about the relationship between lines of force and equipotential surfaces?

They are orthogonal, i.e. they intersect at right angles

... as can be seen from diagram (b) in the previous frame.

Now move on to the next frame and decide which of the ten statements are true.

51

(a) The p.d. between two points is the work in Joules which must be done in moving a charge of +1 C from the point of lower potential to that of higher.
(b) The p.d. in (a) is dependent upon the path taken between the two points.
(c) The unit of potential difference is the volt per metre.
(d) The electric field strength in any direction is the rate at which the potential decreases in that direction.
(e) The Newton per Coulomb equals the volt per metre.
(f) An electric dipole is a pair of equal charges separated by a small distance.
(g) The potential at a point due to an electric dipole is inversely proportional to the square of the distance from the point to the dipole.
(h) Electric fields are directed along equipotential surfaces.
(i) Electric field strength is equal to potential gradient.
(j) The potential at a point in an electric field may be defined as the work required to bring a unit positive charge from infinity to the point.

52

$$\boxed{\text{True: a, d, e, g, j}}$$

(b) Potential difference is a scalar quantity. (Frame 39)
(c) The volt per metre is the unit of potential gradient. (Frame 43)
(f) This definition is insufficient. (Frame 47)
(h) The E-vector is at right angles to equipotential surfaces. (Frame 49)
(i) Don't forget the minus sign. (Frame 42)

53

SUMMARY OF FRAMES 34–52

Quantities and units

Quantity	Symbol	Unit	Abbreviation
Electric potential difference	V	Volt	V
Electric potential	V	Volt	V
Electric potential gradient	V/m	Volt per metre	$V\,m^{-1}$
Electric moment	m	Coulomb-metre	C m

DEFINITIONS

The *potential difference* between two points is the work required to move a unit positive charge from the point of lower potential to that of higher.

An *electric dipole* is a pair of equal and opposite charges separated by a distance which is small compared with the distance to the point at which we wish to measure its effects.

An *equipotential surface* is one at all points on which the potential is the same; it is at right angles to lines of force.

The *potential gradient* in any direction in an electric field is the rate at which the potential falls off in that direction.

EQUATIONS

$$V_B - V_A = dV - \int E \cos \theta \, dl; \qquad E_x = -\frac{dV}{dx}$$

54

EXERCISES

(i)   Find the electric potential at a point distant 10 cm from a point charge of 20 nC in air.

(ii) An electric field is produced by three point charges, each of 15 nC placed 10 cm, 20 cm, and 30 cm along the X-axis. Calculate the potential at a point 10 cm on the positive Y-axis. All distances are from the origin.

(iii) A very long cylindrical wire of negligible radius carries a charge of 2 μC per metre length in air. Calculate the electric potential at a point 35 cm vertically above it.

(iv) Draw the field pattern due to an isolated point charge. Superimpose on the diagram the potential map (equipotential surfaces). Take the charge to be 50 nC and consider distances of 10 cm, 20 cm, 30 cm, 40 cm and 50 cm from it.

   On your diagram the equipotential surfaces will appear as circles though in reality, of course, they are concentric spheres.

(v)   Calculate the electric potential at a point specified by the polar co-ordinates 0.2 $\angle$ 30° m from the centre of a dipole of moment 15 pC m.

(vi) Is it possible that the charge of the dipole in (v) could be 75 pC?

55

SHORT EXERCISES ON PROGRAMME 2

The numbers on the right refer to the frame where the answer may be found.

1. What is meant by the charge on a body? (3)
2. State Coulomb's law for the force between point charges. (4)
3. What is the nature of the force between like charges? (4)
4. Define electric field strength and give its unit. (10 & 11)
5. State the relationship between E, Q and F. (11)
6. What is a line of force? (12)
7. What is the unit of electric flux density? (27)
8. State Gauss's theorem. (25)
9. State the relationship between E, ε_0 and D. (19)
10. Define the electric potential difference between two points. (36)
11. What is the unit of potential difference? (37)
12. What is the unit of electric potential gradient? (43)
13. Define the electric potential at a point. (46)
14. Define an electric dipole. (47)
15. What is the unit of electric moment? (48)
16. What is an equipotential surface? (49)

56

ANSWERS TO EXERCISES

Frame 18

(i) 1.6×10^{-18} C; (ii) 3.125×10^{18}; (iii) (a) 22.47 kN C^{-1} acting due East (i.e. along the positive X-axis), (b) 9.987 kN C^{-1} acting due South; (iv) 52.32 μN acting in a direction South 15.52° East (assuming that A is at the top left hand corner of the rectangle); (v) 103.2 kN C^{-1} acting along $\overrightarrow{BA}$.

Frame 35

(ii) 8.345×10^{-7} μC; (iii) for example, at $x = 15$ cm, $E = 0.199 \times 10^8$ N C^{-1}.

Frame 54

(i) 1798 V; (ii) 1982 V; (iii) 37.75 kV; (iv) at the distances specified, the potentials are respectively 4494 V, 2246 V, 1498 V, 1123 V, 899 V; (v) 2.918 V; (vi) no, because then $2l = 0.2$ m which is the same as the distance (d) to the point where the potential is being determined; for a dipole, $d \gg 2l$.

Programme 3

ELECTROSTATIC FIELDS 2

1

INTRODUCTION

In this programme we continue the study of charges at rest and consider:

- conductors in electrostatic fields
- capacitance and capacitors
- dielectrics
- insulators in electrostatic fields, polarisation
- electric images
- energy stored in electric fields
- mechanical forces in electric fields
- conditions at a boundary between different media in electric fields.

After you have studied this programme you should be able to:

- calculate the capacitance of various types of capacitor including those comprising parallel plates, concentric spheres, coaxial cables and parallel wires
- calculate the capacitance of capacitors having mixed dielectrics
- calculate the equivalent capacitance of a number of capacitors connected in series or parallel or series/parallel
- describe how the capacitance of a parallel plate capacitor varies with its plate area and separation
- decide on the optimum dimensions of a coaxial cable for economy of insulating material
- calculate the capacitance of a system of conductors situated above earth
- calculate the energy stored in an electric field in terms of capacitance and charge and in terms of the electric field vectors E and D
- calculate the mechanical forces exerted in electric fields
- describe and calculate the way in which lines of force are refracted as they pass from one type of medium to another in an electric field
- describe how polarization takes place in a dielectric
- explain how the method of electric images can be used to calculate the capacitance of systems of conductors in the presence of earthed conducting planes

2

CONDUCTORS IN ELECTROSTATIC FIELDS

A good conductor is a material within which electric charges can move freely. There are good conductors to be found among solids, liquids and gases. In metallic conductors (for example, copper or aluminium) electrons can move about inside very easily but cannot move away from the surface and out of the metal.

As stated in Frame 1 of Programme 2, charges are at rest in electrostatic fields. There can be no electric field inside a conductor under electrostatic conditions, therefore, otherwise electrons would move. When a conducting body is placed in an electrostatic field there is a momentary movement of charge inside it so as to cancel out the original field now occupied by the conductor. The diagrams show the effect of placing (a) a rectangular and (b) a spherical conductor in an electrostatic field.

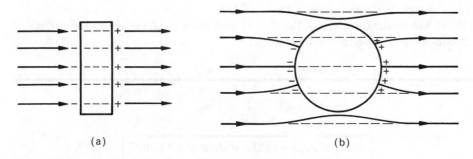

(a)   (b)

The original field is shown dashed and charges appear on the surface of the conductors. You will see that, in the case of the spherical conductor, the field is distorted because the lines of force must enter it at right angles. There can be no E field inside the sphere for the reason mentioned above. Similarly there can be no component of an E field vector tangential to the surface of the sphere otherwise electrons would move along it. The consequence of this is that there can be no difference of potential (i.e. no potential gradient) inside or on the surface of the sphere. This is because, as we saw in Programme 2, Frame 42, $E = -dV/dx$. If E is zero, therefore, the potential gradient, dV/dx, is also zero.

The surface of the sphere or any conducting body is therefore an equipotential surface and everywhere inside the conductor the potential is the same as that at the surface. Electric lines of force and equipotentials intersect at right angles which is why the field in diagram (b) above is distorted.

Now apply Gauss's theorem to any closed surface within a charged spherical conducting body and deduce where the charge must be.

3

> The charge resides entirely on the outer surface of the conductor

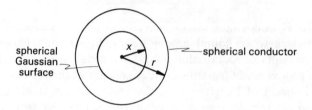

Let the radius of the sphere be r and take any closed spherical surface of radius x ($<r$) as the Gaussian surface.

Because $E = 0$ then $D = 0$ and so $\psi = 0$.

By Gauss's theorem, therefore, the flux crossing the Gaussian surface is zero. Since this is true for *any* surface of radius less than r, there is no charge inside the conductor and its charge must exist entirely on its surface.

What will be the direction of the E- and D-vectors at all points on the surface of a conductor under electrostatic conditions?

4

> At right angles to the surface at all points

... because the surface is an equipotential surface.

The charge per unit area of surface is called *surface charge density* whose symbol is the lower case (small) Greek letter σ (sigma) and for which the unit is the Coulomb per square metre (C m^{-2}).

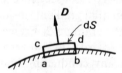

The conductor shown in the diagram has a surface charge density of σ. The Gaussian surface of area dS encloses a small part of the surface of the conductor. The sides ac and bd are at right angles to the surface.

Give an expression for the flux crossing the Gaussian surface and state where it crosses the surface.

5

$\sigma \, dS$ and all of it crosses the outer surface cd

The charge enclosed by the Gaussian surface is $\sigma \, dS$ and so by Gauss's theorem the flux out of the surface abcd is equal to $\sigma \, dS$.

The flux crossing the inner surface ab is zero because the flux density D is zero inside the conductor. The flux crossing the sides ac and bd is zero because D is parallel to them (lines of force are at right angles to equipotential surfaces). It follows that all of the flux must cross cd and that the flux density D immediately outside the conductor is equal to the surface charge density (i.e. $D = \sigma$).

What is the relationship between electric field strength and surface charge density?

6

$$E = \sigma / \varepsilon$$

... since $D = \sigma$ and $E = D / \varepsilon$.

CAPACITANCE

If we take an isolated conductor of any shape whatever and place a charge of Q coulomb on it then its electric potential will rise. Let us assume that it rises to V volt. If, now, we place another charge of Q coulomb on it then by the Principle of Superposition its potential will rise to $2V$ volt.

As charge is added so the potential will rise proportionately, and we can say that $Q \propto V$.

Using a constant of proportionality we write $Q = CV$.

C is called the capacitance of the isolated conductor and is a measure of its capacity for storing charge.

What is the unit of capacitance?

7

> Coulomb per volt ($C V^{-1}$). This is called the Farad.

We now consider two conductors, again of any shape whatever, and initially uncharged. If a charge of Q is removed from one of the conductors and passed to the other, an electric field will be set up between and around the conductors and an electric potential difference will exist between them. Let us call this V volt.

If n charges, each of Q coulomb, are taken from one of the conductors and placed on the other then the potential difference between them will be nV volt by the Principle of Superposition.

As before, we have $Q \propto V$ and again we introduce a constant of proportionality, C, such that $Q = CV$. The constant C is called the capacitance between the conductors and an arrangement of conductors which has capacitance is called a capacitor. The conductors themselves are called plates or electrodes.

The circuit symbol for a capacitor is as shown above although it may consist of parallel plates, concentric cylinders, concentric spheres or any other arrangement of conducting surfaces.

Which of the following do you think capacitance depends on: surface area of the plates; type and/or thickness of the material between the plates?

8

> The capacitance of a capacitor depends upon all of these.

The space between the plates of a capacitor may be a vacuum or air or some other material. Ideally, this material should be one in which charges do not move and such materials are called insulators. Faraday gave the name *dielectric* to the insulating material between the plates of a capacitor and it is found that for a given arrangement of plates the capacitance is greater with a dielectric than it is with a vacuum between the plates by a factor which is a constant for the dielectric.

This constant is known as the *dielectric constant* or the *relative permittivity* of the material of the dielectric. Its symbol is ε_r and some typical values are as follows:

Dielectric material	Relative permittivity
A vacuum	1 (by definition)
Dry air at NTP	1.000 59
Dry paper	1.5–4
Polythene	2.3
Glass	6
Water	81

A capacitor has a capacitance of 10 000 pF when the dielectric is glass. What would its capacitance be if the glass were replaced with air?

9

$$1667 \text{ pF } (10\,000 \times (1.000\,59/6))$$

We have now met three different symbols associated with permittivity:

ε the absolute permittivity of a field medium
ε_0 the absolute permittivity of free space
ε_r the relative permittivity of a medium.

We will now obtain a relationship between them.

(a)   (b)

The two capacitors shown have identical geometry and the same potential difference, V, is maintained between both pairs of plates. Between the plates, capacitor (a) has a vacuum while (b) has a material of permittivity ε.

Under these conditions the electric field strength, E, will have the same value at corresponding points in each capacitor.

For capacitor C_1, $D_1 = \varepsilon_0 E$ where D_1 is the flux density at a point in the vacuum.

For capacitor C_2, $D_2 = \varepsilon E$ where D_2 is the flux density at the corresponding point in its dielectric.

Now obtain an expression relating Q_1, Q_2, ε_0 and ε.

10

$$\boxed{Q_2/Q_1 = \varepsilon/\varepsilon_0}$$

Since $\boldsymbol{D} = \sigma = Q \times$ plate area A (Frame 5), then $\boldsymbol{D} \propto Q$ for a given A so $Q_2/Q_1 = \boldsymbol{D}_2/\boldsymbol{D}_1 = \varepsilon\boldsymbol{E}/\varepsilon_0\boldsymbol{E} = \varepsilon/\varepsilon_0$.

We have also that $Q \propto C$ (V is the same for both capacitors) therefore $C_2/C_1 = Q_2/Q_1 = \varepsilon/\varepsilon_0$. As stated in Frame 8, this ratio is called the relative permittivity of the dielectric (ε_r). Thus $\varepsilon = \varepsilon_0\varepsilon_r$.

Using the data given in Frame 8 and taking $\varepsilon_0 = 8.854 \times 10^{-12}$ SI units (we shall find out what the SI unit of permittivity is later), calculate the absolute permittivity of (a) glass, (b) water.

11

$$\boxed{\text{(a) } 53.12 \times 10^{-12} \text{ SI units; (b) } 717 \times 10^{-12} \text{ SI units}}$$

INSULATORS IN ELECTROSTATIC FIELDS

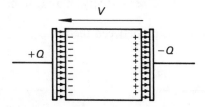

When a slab of dielectric is inserted between charged plates, the electrons of the atoms of the dielectric material experience a force pulling them towards the positive plate of the capacitor while the protons are pulled towards the negative plate. Since electrons cannot move through an insulator, each atom becomes like a dipole. The net effect within the dielectric material is nil but near the junction with the plates there are surface layers of charge, negative (electrons) at the positive plate and positive near the negative plate. Polarisation is said to have taken place and the polarising field is opposed at the surface of the dielectric material. The effect is to reduce the electric field strength inside the dielectric which means a lower potential difference between the plates. Since the charge is unchanged, the capacitance must rise. Compare this with a conductor in which the original field is completely eliminated within it (see Frame 2).

CALCULATION OF CAPACITANCE

To calculate the capacitance of various arrangements of conductors we use the relationship $C = Q/V$ which means, in general, obtaining an expression for the potential difference between the conductors and dividing this into the charge Q.

Example: the capacitance formed by parallel plates

The capacitor shown has plates of area A separated by a dielectric of relative permittivity ε_r. The p.d. maintained between the plates is V and the charge on each plate is Q with a charge density σ. We assume that the field between the plates is uniform and that it is negligible elsewhere.

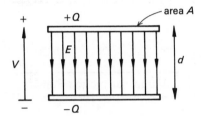

The potential difference between the plates is given by

$$V = -\int_L E \cos \theta \, dl \qquad \text{(see Frame 37 of Programme 2)}$$

We take one of the lines of force as the path L and note that along this path E is uniform and $\cos \theta = 1$ so that

$$V = -E \int_L dl$$

The result of adding all the dl terms which make up the path L is d, the distance between the plates.

Therefore $V = Ed$ with the top plate at the higher potential.

The flux density immediately outside the conductors and between the plates is given by $D = \sigma$ (Frame 5).

Now $E = D/\varepsilon = \sigma/\varepsilon = \sigma/\varepsilon_0 \varepsilon_r$.

therefore $V = Ed = \sigma d/\varepsilon_0 \varepsilon_r$.

Now obtain an expression for the capacitance of the parallel plate arrangement remembering that $C = Q/V$ and that $Q = \sigma \times$ plate area.

13

$$C = \frac{A\varepsilon_0\varepsilon_r}{d} \text{ (F)}$$

$$C = \frac{Q}{V} = \frac{(\sigma A)}{V} = \frac{\sigma A\varepsilon_0\varepsilon_r}{\sigma d} = \frac{A\varepsilon_0\varepsilon_r}{d} \text{ (F)}$$

What is the unit of ε_0?

14

Farad per metre (F m^{-1})

Transposing the formula we see that the unit of ε_0 is the unit of C times the unit of d divided by the unit of A. Remember, ε_r has no units.

The expression derived above for the capacitance of a parallel plate capacitor assumes that the lines of force are uniform between the plates and non-existent elsewhere. In practice there will be a 'fringing flux' present as shown in the diagram. The effect is that the charge density is higher near the edges and the capacitance is higher than that given by the formula. However, for capacitors whose plate length is large compared with their separation, fringing effects are negligibly small and the formula gives sufficiently accurate results.

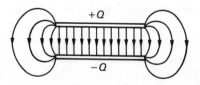

How may the capacitance of a parallel plate capacitor be increased?

15

> Since $C = A\varepsilon/d$, we can:
> (a) increase ε; (b) decrease d; (c) increase A.

(a) We have seen that capacitance is increased when a dielectric material is inserted between the plates by the process of polarisation. However, there is a limit to the range of values of ε_r available in suitable dielectric materials.

(b) The separation cannot be reduced too much otherwise the dielectric material becomes too fragile and is liable to break during manufacture. In addition the dielectric must be thick enough to withstand the working voltage V of the capacitor.

(c) There is clearly a limit to the physical size of a capacitor. Too large an area would make it unacceptably unwieldy.

It might be thought that by using multiple capacitors we could obtain a higher capacitance. Let us see if this is the case.

16

First we shall try connecting two capacitors in series as shown below.

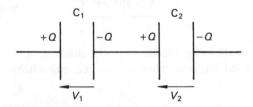

The application of a potential difference will produce a charge $+Q$ on the left-hand plate of C_1 and a charge of $-Q$ on the right-hand plate of C_2. The charge $+Q$ on the left-hand plate of C_1 will attract electrons to the facing plate, giving it a charge of $-Q$. Similarly, the charge $-Q$ on the right-hand plate of C_2 will repel electrons from the facing plate, giving it a charge $+Q$. Note that the two inner plates and the connecting wire can be considered as a conductor in an electric field and that the charge distribution agrees with that discussed in Frame 2.

A single capacitor having a capacitance equivalent to that of the series combination would have a charge of Q coulomb and a potential difference of $(V_1 + V_2)$ volts between its plates. Let this equivalent capacitance be C_T.

See if you can obtain an expression relating C_T, C_1 and C_2.

17

$$1/C_T = 1/C_1 + 1/C_2$$

Since the equivalent capacitance is C_T

then $C_T = Q/V_1 + V_2$

and $Q/C_T = V_1 + V_2 = Q/C_1 + Q/C_2$.

Dividing throughout by Q gives the answer in the box.

In general, for n capacitors connected in series the equivalent capacitance, C_T is given by $\dfrac{1}{C_T} = \sum_{a=1}^{n} \dfrac{1}{C_a}$.

Three capacitors having capacitances of 10 μF, 20 μF and 30 μF are connected in series. Calculate the equivalent capacitance of the series combination.

18

$$5.455 \ \mu F$$

So this gives us a smaller capacitance than the smallest of the three capacitors.

Let us now see what happens when we connect capacitors in parallel.

The capacitors will have the same potential difference across their plates but the charge on each capacitor will be different. A single capacitor of equivalent capacitance will have the same potential difference between its plates and the same total charge on its plates.

Obtain an expression for the capacitance of this equivalent capacitor.

19

$$\boxed{C_T = C_1 + C_2}$$

For the 'equivalent' capacitor, $C_T = (Q_1 + Q_2)/V$

$$\therefore C_T V = Q_1 + Q_2 = C_1 V + C_2 V$$

Dividing by V gives the answer in the box.

So, to find the equivalent capacitance of a number of capacitors connected in parallel we simply add their individual capacitances. *Connecting capacitors in parallel is therefore a way of achieving a higher capacitance.*

20

Example

Two identical parallel plate capacitors are connected in parallel to form a capacitance of 600 pF. If the plates measure 20 cm × 30 cm and they are separated by a distance of 5 mm, calculate the relative permittivity of the dielectric material.

Solution

The first step is to find the capacitance of each capacitor. Since they are connected in parallel we use $C_T = C_1 + C_2$. But $C_1 = C_2$ so that $C_T = 2C_1$ and $C_1 = 300$ pF.

Next we use the expression for the capacitance of a parallel plate capacitor and transpose it to make ε_r the subject.

$$C = \frac{A \varepsilon_0 \varepsilon_r}{d}$$

$$\therefore \varepsilon_r = \frac{Cd}{A\varepsilon_0} = \frac{300 \times 10^{-12} \times 5 \times 10^{-3}}{0.3 \times 0.2 \times 8.854 \times 10^{-12}} = \underline{2.82}$$

(a) From this result and using the table in Frame 8, suggest what the dielectric material might be.
(b) What would be the equivalent capacitance if the capacitors were connected in series?

21

> (a) Dry paper; (b) 150 pF

Which of the following statements are true?

(a) Only metals are good conductors because only they have 'free' electrons which move easily.

(b) There can be no electric field strength (E-vector) inside a conductor under electrostatic conditions.

(c) Conductors are surfaces of equipotential in electrostatic fields.

(d) Lines of force are tangential to the surface of cylindrical conductors in electrostatic fields.

(e) Charge cannot exist in or on a conductor in a truly electrostatic field.

(f) The unit of surface charge density is the coulomb per square metre.

(g) Immediately outside the surface of a conductor in an electrostatic field the electric flux density is equal to the surface charge density.

(h) Capacitance is directly proportional to charge.

(i) For a given charge, capacitance is independent of potential difference.

(j) The unit of relative permittivity is the Farad per metre.

(k) The capacitance of a parallel plate capacitor is inversely proportional to the plate separation.

(l) The equivalent capacitance of two capacitors having capacitance of C_1 farad and C_2 farad when they are connected in series is given by $1/C_1 + 1/C_2$.

(m) Capacitance is measured in coulomb per volt.

(n) The effect of polarisation in a dielectric is to produce an electric field in opposition to the polarising field.

(o) The absolute permittivity of a dielectric material whose relative permittivity is ε_r is given by $\varepsilon = \varepsilon_r / \varepsilon_0$.

(p) The equivalent capacitance of two equal capacitors when they are connected in series is half the capacitance of one of them.

22

> True: b, c, f, g, h, k, m, n, p

(a) There are liquids, gases and non-metallic solids which are also good conductors.

(Frame 2)

(d) They are normal to such surfaces.   (Frame 2)
(e) Charge can exist on the surface.   (Frame 3)
(i) For a given charge, $C \propto 1/V$.   (Frame 6)
(j) Relative permittivity has no unit, it is a simple ratio.   (Frame 8)
(l) It is the reciprocal of this.   (Frame 17)
(o) Absolute permittivity, $\varepsilon = \varepsilon_0 \varepsilon_r$.   (Frame 10)

23

SUMMARY OF FRAMES 1–22

Quantities and units

Quantity	Symbol	Unit	Abbreviation
Surface charge density	σ	Coulomb per metre^2	C m^{-2}
Capacitance	C	Farad	F
Permittivity	ε	Farad per metre	F m^{-1}
Dielectric constant ⎫ Relative permittivity ⎬	ε_r	—	—
Dielectric strength ⎫ Electric stress ⎬	E	Volt per metre	V m^{-1}

Formulae

$D = \sigma$ immediately outside the surface of a conductor

$$C = \frac{Q}{V}$$

$$\varepsilon = \varepsilon_0 \varepsilon_r$$

$$C = \frac{A \varepsilon_0 \varepsilon_r}{d} \text{ for a parallel plate capacitor}$$

$$C_T = C_1 + C_2 + \ldots C_n \text{ for } n \text{ capacitors connected in parallel}$$

$$\frac{1}{C_T} = \frac{1}{C_1} + \frac{1}{C_2} + \ldots + \frac{1}{C_n} \text{ for } n \text{ capacitors connected in series}$$

24

EXERCISES

(i) Calculate the absolute permittivity of paraffin whose relative permittivity is 2.1.

(ii) A certain dielectric material has an absolute permittivity of 40×10^{-12} F m^{-1}. Calculate its relative permittivity.

(iii) A conductor is charged to 300 V and its capacitance is then 200 pF. What is the charge on the conductor?

(iv) Two conductors having charges of ± 25 μC on them form a capacitor of capacitance 15 μF. Calculate the potential difference between them.

(v) A capacitor is made from two parallel circular plates 3 cm in diameter separated by a slab of dielectric of similar area and 5 mm thick. Calculate the capacitance of the capacitor. The relative permittivity of the dielectric is 3.

(vi) A parallel plate capacitor has plates 50 cm square and 5 cm apart. There is a potential difference of 20 kV between the plates. Calculate the charge on each plate if they are separated by air.

(vii) The space between the plates of the capacitor of the previous question is filled with a dielectric material having relative permittivity of 8. Calculate the capacitance.

(viii) Two parallel plates of area 100 cm^2 each are separated by a sheet of mica 0.1 mm thick. The relative permittivity of the mica is 4. The potential difference between the plates is 500 V.
Calculate: (a) the capacitance of the capacitor; (b) the charge on the plates; (c) the electric flux density in the mica; (d) the electric field strength in the mica.

(ix) Give all possible values of capacitance available from three capacitors of capacitance 3 μF, 9 μF and 27 μF.

(x) Two capacitors C_1 and C_2 are connected in series across a potential difference of V volt. The potential difference between the plates of C_1 and C_2 are respectively V_1 and V_2. Show that

$$V_2 = \frac{V C_1}{C_1 + C_2}$$

25

THE CAPACITANCE BETWEEN TWO VERY LONG CONCENTRIC CYLINDERS

This is important because coaxial cables may be thought of as comprising concentric cylinders. The diagram shows the arrangement.

 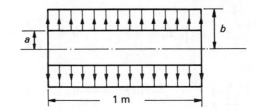

The inner cylinder is charged to q coulomb per metre length. To find the capacitance we use $C = q/V$ and the answer is in farad per metre.

The potential difference between the cylinders, $V = -\int E_r \cos \theta \, dr$ where E_r is the field strength at some point within the dielectric and is given by D_r/ε, and $D_r =$ the flux/the area over which it exists. Now the flux (ψ) is related to q by Gauss's theorem and to apply this we need a Gaussian surface. Suggest a suitable one.

26

A cylinder of radius r $(b > r > a)$ and length 1 m

All of the flux crosses the curved surface of this Gaussian surface and is equal to the charge (q) enclosed by it.

$$D = \frac{\psi}{2\pi r \times 1} = \frac{q}{2\pi r}$$

$$E = \frac{D}{\varepsilon} = \frac{q}{2\pi r \varepsilon}$$

$$V = -\int_L \frac{q}{2\pi r} \, dr$$

Suggest a path over which to perform the integration.

27

> One of the lines of force between the cylinders

Call this path L. The limits of the path are $r = b$ and $r = a$ and since the electric field strength is along the path at all points then $\theta = 0$ and $\cos \theta = 1$.

Thus
$$V = -\int_b^a E_r \, dr = +\int_a^b E_r \, dr = \int_a^b \frac{D_r}{\varepsilon} \, dr = \int_a^b \frac{q}{2\pi\varepsilon r} \, dr \quad \text{(V)}$$

$$= \frac{q}{2\pi\varepsilon} \int_a^b \frac{dr}{r} = \frac{q}{2\pi\varepsilon} [\ln r]_a^b = \frac{q}{2\pi\varepsilon} \ln \frac{b}{a} \quad \text{(V)}$$

Now complete the problem by writing down the expression for the capacitance between the cylinders.

28

$$\boxed{C = \frac{q}{V} = \frac{2\pi\varepsilon}{\ln(b/a)} \quad \text{(F m}^{-1}\text{)}}$$

We have that
$$E_r = \frac{q}{2\pi\varepsilon r} \quad \text{(V m}^{-1}\text{)}$$

and that
$$V = \frac{q}{2\pi\varepsilon} \ln \frac{b}{a}$$

from which we see that

$$rE_r = \frac{V}{\ln(b/a)} \quad \text{and} \quad \underline{E_r = V/(r \ln(b/a))} \quad \text{(V m}^{-1}\text{)}$$

This is an expression for the electric field strength (also known as electric stress) at points in the dielectric material at radius r from the axis. Where in the dielectric will the maximum and minimum stress (E_{max} and E_{min}) occur?

E_{max} occurs at the surface of the inner conductor ($r = a$).
E_{min} occurs at the outer cylinder where $r = b$.

For economy of dielectric material it is important to keep the value of E_{max} as small as possible. The stress at all other points in the dielectric will then also be reduced.

Let's see how we can achieve this.

We have that
$$E_{max} = \frac{V}{a \ln(b/a)} \quad \text{(V m}^{-1}\text{)}$$

For a given working voltage (V) we need to make the denominator of this expression as large as possible in order to minimise E_{max}. The overall diameter of the cable is $2b$ and assuming this to be fixed, then the variable is a (the radius of the inner conductor).

To find the condition for a maximum value we differentiate the denominator with respect to a and equate to zero.

Do this now and find the condition for E_{max} to be as small as possible.

30

The condition is that $b/a = e$ (2.718)

Here is the working (remember, b is a constant):

$$\frac{d}{da}\left[a \ln \frac{b}{a} \right] = \frac{d}{da}[a(\ln b - \ln a)]$$

$$= 1(\ln b - \ln a) + a(-1/a)$$

$$= (\ln b - \ln a) - 1 \qquad \text{(i)}$$

For a maximum or minimum value this must equal 0

i.e. $\qquad \ln(b/a) - 1 = 0 \Rightarrow \ln(b/a) = 1 \qquad$ and $\qquad \underline{b/a = e}$.

Differentiating (i) gives $(-1/a)$ which is negative so we have the condition for $(a \ln(b/a))$ to be a maximum and therefore E_{max} to be a minimum.

31

Example

A coaxial cable has a conductor (inner cylinder) of radius 2.5 cm and a sheath (outer cylinder) of radius 4 cm. The space between the cylinders is filled with a dielectric material having a relative permittivity of 4.

Calculate the capacitance per metre of the cable.

For the same sheath dimensions, what radius of conductor would give the least electric stress within the dielectric?

Solution

The capacitance per metre of a coaxial cable (concentric cylinders) is given by

$$C = \frac{2\pi\varepsilon_0\varepsilon_r}{\ln(b/a)} \quad (\text{F m}^{-1})$$

In this case, $\varepsilon_r = 4$, $b = 0.04$ m, $a = 0.025$ m, $\varepsilon_0 = 8.854 \times 10^{-12}$ F m^{-1}. Substituting these values into the formula we get $C = 473$ pF m^{-1}.

For least electric stress within the dielectric material we must make $b/a = e$ so that $a = b/e = 0.04/2.718$ m $= 0.0147$ m $= \underline{1.47 \text{ cm}}$.

Now you try this one:

A coaxial cable has a sheath diameter of 7.6 cm and is to operate at 85 kV. If there is to be the least possible stress at the conductor surface, calculate (i) the radius of the conductor and (ii) the value of the maximum stress.

32

$$\boxed{\text{(i) } 1.4 \text{ cm}; \text{(ii) } 60.7 \text{ kV/cm}}$$

Here is the working: for least stress, the conductor radius (a) must equal the sheath radius (b) divided by e, so that $a = b/e = 3.8/2.718 = 1.4$ cm.

The maximum electric stress is given by $E_{\max} = \dfrac{V}{a\ln(b/a)}$ V m^{-1}.

This is a minimum when $\ln(b/a) = 1$ and is then $V/a = (85/1.4)$ kV/cm.

The electric stress at the surface of the conductor is thus $\underline{60.7 \text{ kV cm}^{-1}}$.

THE CAPACITANCE BETWEEN PARALLEL CYLINDERS

Important examples of this arrangement are parallel wires such as overhead transmission lines.

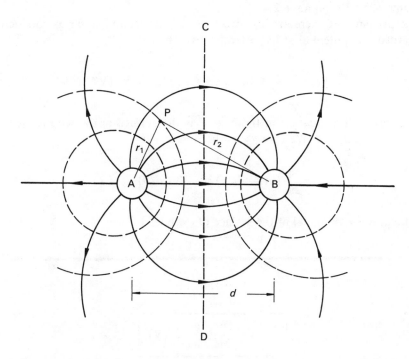

The radius of each cylinder is a and the distance between their centres is d where $d \gg a$.

The cylinders (A and B) are charged to $+q$ Coulomb and $-q$ Coulomb respectively, the resulting lines of force being shown (solid lines) and the equipotential surfaces being drawn everywhere at right angles to them.

The cylinders are considered to be isolated from any other earthed conductor or earthed conducting plane.

The point P may be ignored for the present. It is referred to at the end of Frame 35. The associated distances r_1 and r_2, together with the symbols C and D are not required until Frames 36 and 37.

The technique for obtaining an expression for the capacitance between the cylinders is to obtain, first, an expression for the potential difference between them and then divide this into the charge per unit length (q). This will give the capacitance per metre length of the 'twin line' as the arrangement is called.

Can you remember the expression for the potential at a point P due to a number of long, straight, parallel line charges?

34

$$V_P = -\frac{1}{2\pi\varepsilon} \sum_{a=1}^{n} q_a \ln d_a \qquad \text{(V)}$$

See Frame 49 of Programme 2.

In the present case there are just two lines in parallel and so far as conductor A is concerned, the potential at its surface is given by

$$V_A = -\frac{1}{2\pi\varepsilon}[+q \ln a + (-q) \ln d] \qquad \text{(V)}$$

Relating this to the equation in the box: P is the surface of the wire A, $q_1 = +q$, $d_1 = a$, $q_2 = -q$, $d_2 = d$. We take $d_2 = d$ because $d \gg a$.

So
$$V_A = \frac{q}{2\pi\varepsilon}(\ln d - \ln a) = \frac{q}{2\pi\varepsilon}\ln(d/a) \qquad \text{(V)}$$

What, then, is the potential at the surface of cylinder B?

35

$$V_B = -\frac{q}{2\pi\varepsilon}\ln\frac{d}{a} \qquad \text{(V)}$$

In this case P is the surface of conductor B, $q_1 = +q$, $d_1 = d$, $q_2 = -q$ and $d_2 = a$.

Thus
$$V_B = -\frac{1}{2\pi\varepsilon}(q \ln d + (-q) \ln a) = -\frac{1}{2\pi\varepsilon}\ln\frac{d}{a} \qquad \text{(V)}$$

The potential difference between the wires, $V_A - V_B = 2\frac{q}{2\pi\varepsilon}\ln\frac{d}{a}$.

The capacitance between the wires is given by $C = q/V$

so that
$$C = \frac{\pi\varepsilon}{\ln(d/a)} \qquad \text{(F m}^{-1}\text{)}.$$

Write down an expression for the potential at any point such as P in the diagram in Frame 33.

36

$$V_P = \frac{q}{2\pi\varepsilon} \ln \frac{r_2}{r_1} \quad \text{(V)}$$

... because $\quad V_P = -\dfrac{1}{2\pi\varepsilon}(q \ln r_1 + (-q) \ln r_2) = \dfrac{q}{2\pi\varepsilon}(\ln r_2 - \ln r_1).$

On the line CD which is the perpendicular bisector of the line joining A and B $r_1 = r_2$, so for any point P on this line $\ln(r_2/r_1) = 0$ and $V_P = 0$. The line CD is the zero equipotential.

The potential at any point on any other equipotential is constant so that r_2/r_1 remains constant as P moves along it. It can be shown that the locus of V_P as it moves along an equipotential surface is a circle with its centre on the line AB.

What can you deduce about the lines of force?

37

They are circles with their centres on CD

Example

Calculate the capacitance per metre of a twin line separated by a horizontal distance of 2 m. The conductors each have a radius of 1.2 cm.

Solution

The capacitance of a twin line is given by

$$C = \frac{\pi\varepsilon}{\ln(d/a)} (\text{F m}^{-1})$$

where a is the conductor radius and d is their separation (Frame 35). In this case $d = 2$ m, $a = 0.012$ m, $\varepsilon_0 = 8.854 \times 10^{-12}$ F m^{-1} (air) so that, substituting these values into the formula, $C = 5.4$ pF m^{-1}.

The result would have been different had we taken into account the fact that lines are usually situated above earth which can be considered to be an equipotential surface. The next frame describes a technique for dealing with this.

38

ELECTRIC IMAGES

The use of the concept of electric images enables us to take into account the effect of earthed planes in capacitance calculations. To see what is meant by an electric 'image', we will consider again the parallel wires arrangement (Frame 33). The diagram is given below.

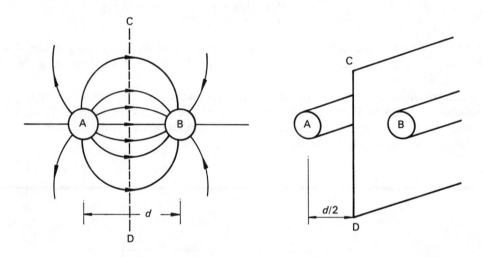

We saw that CD is an equipotential surface. Now there is no component of electric field along such a surface so that a conductor may be placed on it without affecting the field pattern in any way. Notice that the lines of force cross it at right angles. Charges of $+q$ coulomb per metre appear on the right-hand side of the conducting surface and $-q$ coulomb per metre appear on the left-hand side. The conductor is said to 'screen' the two halves of the field from one another. If the wire B is now removed the field between wire A and the conducting plane CD remains unchanged. Wire B is called the negative image of wire A in the plane CD.

What, then, is the capacitance between wire A and the conductor at CD?

39

> The potential difference between the wire and the conducting plane is one-half of that between the two wires, and since $C \propto 1/V$ then the capacitance is doubled so $C = 2\pi\varepsilon/\ln(d/a)$ (F m^{-1}).

EXAMPLE OF THE USE OF ELECTRIC IMAGES

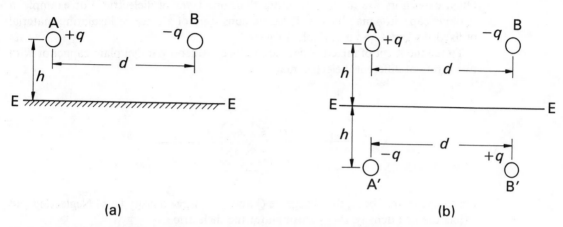

(a)   (b)

The diagram on the left shows a twin line having conductors of radius a, charged to q coulomb per metre length and separated by a horizontal distance of d metre at a height h metre above level earth. The field pattern between the conductors and the earth is the same as that between the conductors and the surface EE when image conductors are introduced as shown in (b). Therefore if we find the capacitance between the wires in the 'image' system of diagram (b) it will be the same as that for the real system of (a).

The potential at the surface of wire A (diagram (b)) is given by

$$V_A = -\frac{1}{2\pi\varepsilon}\Sigma q \ln r = -\frac{1}{2\pi\varepsilon}[+q \ln a + (-q) \ln d + (-q) \ln(2h) + q \ln(d^2 + (2h)^2)^{1/2}]$$

$$= +\frac{q}{2\pi\varepsilon}[\ln d + \ln(2h) - \ln a - \ln(d^2 + (2h)^2)^{1/2}] = \frac{q}{2\pi\varepsilon}\ln\left[\frac{d2h}{a(d^2 + (2h)^2)^{1/2}}\right]$$

Similarly
$$V_B = -\frac{q}{2\pi\varepsilon}\ln\left[\frac{d2h}{a(d^2 + (2h)^2)^{1/2}}\right] = -V_A$$

$$\therefore V_A - V_B = 2V_A = \frac{q}{\pi\varepsilon}\ln\left[\frac{d2h}{a(d^2 + (2h)^2)^{1/2}}\right] \quad \text{(V)}$$

The capacitance between the wires is

$$\frac{q}{V_A - V_B} = \frac{\pi\varepsilon}{\ln\left[\dfrac{2hd}{a(d^2 + (2h)^2)^{1/2}}\right]} \quad \text{(F m}^{-1}\text{)}$$

Note that if $h \gg d$ this expression reduces to that for an isolated twin line.

41

CAPACITORS WITH MIXED DIELECTRICS (COMPOSITE CAPACITORS)

Often capacitors are made with more than one layer of dielectric. For example, a variable capacitor may have a dielectric consisting of a piece of insulating material of fixed thickness and a variable air gap.

To see the effect of mixed dielectrics let us consider a parallel plate capacitor with two layers of different dielectric material.

Let the plate area be A, the charge be Q and the charge density be σ. Neglecting end effects the flux density $D = \sigma$ throughout the dielectric.

$\therefore$ for layer 1: $\quad E_1 = D/\varepsilon_0\varepsilon_{r1} = \sigma/\varepsilon_0\varepsilon_{r1}$ (i)

and for layer 2: $\quad E_2 = D/\varepsilon_0\varepsilon_{r2} = \sigma/\varepsilon_0\varepsilon_{r2}$ (ii)

Dividing (i) by (ii) we see that $E_1/E_2 = \varepsilon_{r2}/\varepsilon_{r1}$ which means that the higher stress is in the material having the lower permittivity.

$$V_1 + V_2 = E_1 d_1 + E_2 d_2$$

$$= \frac{\sigma d_1}{\varepsilon_0\varepsilon_{r1}} + \frac{\sigma d_2}{\varepsilon_0\varepsilon_{r2}} = \frac{\sigma}{\varepsilon_0}\left[\frac{d_1}{\varepsilon_{r1}} + \frac{d_2}{\varepsilon_{r2}}\right] = \frac{\sigma}{\varepsilon_0}\left[\frac{\varepsilon_{r2}d_1 + \varepsilon_{r1}d_2}{\varepsilon_{r1}\varepsilon_{r2}}\right]$$

The capacitance $\quad C = \dfrac{Q}{V} = \dfrac{\sigma A}{V} = \dfrac{A\varepsilon_0\varepsilon_{r1}\varepsilon_{r2}}{\varepsilon_{r2}d_1 + \varepsilon_{r1}d_2}$ (F) (iii)

If we connect in series two capacitors, each having the characteristics of one dielectric of the composite capacitor, the equivalent capacitance is given by:

$$\frac{1}{C_T} = \frac{1}{C_1} + \frac{1}{C_2} = \frac{d_1}{A\varepsilon_0\varepsilon_{r1}} + \frac{d_2}{A\varepsilon_0\varepsilon_{r2}}$$

$$= \frac{1}{A\varepsilon_0}\left[\frac{d_1}{\varepsilon_{r1}} + \frac{d_2}{\varepsilon_{r2}}\right] = \frac{1}{A\varepsilon_0}\left[\frac{\varepsilon_{r2}d_1 + \varepsilon_{r1}d_2}{\varepsilon_{r2}\varepsilon_{r1}}\right] \quad (F^{-1})$$

$$\therefore C = \frac{A\varepsilon_0\varepsilon_{r1}\varepsilon_{r2}}{\varepsilon_{r2}d_1 + \varepsilon_{r1}d_2} \quad (F) \tag{iv}$$

Comparing (iii) and (iv) we see that the layers of a composite capacitor are effectively two series connected capacitors.

GRADING OF CABLES

Concentric cables are also often constructed with different layers of dielectric material. Let us see what advantage there is in doing this.

Consider a cable having a conductor of radius 6 mm, two layers of dielectric and an outer conductor of radius 20 mm. Dielectric A has a relative permittivity of 4 and a maximum electric stress of 1.8 MV m^{-1}, while the corresponding values for dielectric B are 2.8 and 2.2 MV m^{-1}.

If the cable is to be operated at maximum voltage, the dielectric material must be changed at a point which will best use their electric field strengths.

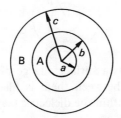

We have that at any radius r: $\qquad E_r = \dfrac{V}{r \ln(b/a)}$ (Frame 28)

For layer A, at radius b: $E_{Ab} = \dfrac{V_A}{b \ln(b/a)} \Rightarrow E_{Ab}b = \dfrac{V_A}{\ln(b/a)}$

For layer A, at radius a: $E_{Aa} = \dfrac{V_A}{a \ln(b/a)} \Rightarrow E_{Aa}a = \dfrac{V_A}{\ln(b/a)}$

Thus $\qquad\qquad\qquad\qquad E_{Ab}b = E_{Aa}a$ (i)

At radius b, the electric flux density in both layers is the same so that $\varepsilon_0\varepsilon_{rA}E_{Ab} = \varepsilon_0\varepsilon_{rB}E_{Bb}$.

Notice that $E_{Ab}/E_{Bb} = \varepsilon_{rB}/\varepsilon_{rA}$ so that to minimise E_A, ε_{rA} must be $> \varepsilon_{rb}$. Also, $E_{Ab} = 2.8 \times 2.2/4 = 1.54$ MV m^{-1}.

It follows from (i) that $b = 1.8 \times 6/1.54 = 7$ mm.

Now $V_{Amax} = a \ln(b/a)E_{Amax} = 6 \ln(7/6)\, 1.8 = 1.66$ kV (MV/m $\times$ mm $=$ kV)

and $V_{Bmax} = b \ln(c/b)E_{Bmax} = 7 \ln(20/7)2.2 = 16.16$ kV.

The maximum working voltage of the cable is thus $1.66 + 16.16 = \underline{17.82\ \text{kV}}$.

Calculate V_{max} if there were a single layer (14 mm thick) of dielectric B only.

43

89

15.98 kV

In this case, $V_{max} = a \ln(c/a) \times E_{max} = 6 \ln(20/6) \times 2.2 = 15.89$ kV. There is therefore considerable benefit to be gained by adding a small layer (1 mm thick) of material A. The maximum working voltage increases by almost 2 kV.

THE CAPACITANCE FORMED BY CONCENTRIC SPHERES

The inner sphere of radius a has a surface charge of Q coulomb. The outer sphere has a radius b and the space between them is filled with a dielectric of relative permittivity ε_r. At some point within the dielectric (say at radius r), the electric stress is given by $E_r = Q/(4\pi\varepsilon_0\varepsilon_r r^2)$ (V m^{-1}).

The potential difference between the spheres is given by

$$V = -\int_b^a E_r \, dr = -\int_b^a \frac{Q}{4\pi\varepsilon_0\varepsilon_r r^2} \, dr = \frac{Q}{4\pi\varepsilon_0\varepsilon_r}\left[\frac{1}{r}\right]_b^a = \frac{Q}{4\pi\varepsilon_0\varepsilon_r}\left[\frac{1}{a} - \frac{1}{b}\right] \quad \text{(V)}$$

The capacitance between the spheres is $Q/V = 4\pi\varepsilon_0\varepsilon_r/((1/a) - (1/b))$.

Rearranging, we obtain finally $C = \dfrac{4\pi\varepsilon_0\varepsilon_r ab}{b - a}$ (F).

Note that if we put $b = \infty$ in the expression for the potential difference between the spheres, we can deduce an expression for the capacitance of an isolated sphere. What do you make it?

44

$C = 4\pi\varepsilon_0\varepsilon_r a$ where a is the radius of the isolated sphere

With $b = \infty$, the potential of the isolated sphere becomes $Q/(4\pi\varepsilon_0\varepsilon_r a)$ and then dividing this into its charge (Q) gives the expression in the box.

45

MULTIPLATE CAPACITORS

A convenient way of increasing capacitance is to increase the effective plate area by using a stack of plates separated by layers of dielectric. Alternate plates are joined together, effectively forming a number of capacitors in parallel. The capacitance of a capacitor having N plates, each of area A and separated by $(N-1)$ layers of similar dielectric material is then given by $(N-1)$ times that of one layer.

Now decide which of the following statements are true:

(a) The electric field strength in the dielectric of a coaxial cable reduces linearly from a maximum at the surface of the outer conductor to a minimum at the surface of the inner conductor.

(b) For a given operating voltage, V, and sheath radius, b, the best economy of dielectric material in a coaxial cable is obtained when the inner conductor radius $a = b/e$ where e is the base of the natural logarithm.

(c) An electric image conductor is of the same sign as the real conductor.

(d) The method of images can be used to calculate the capacitance of conductor systems in the presence of an earthed conducting plane.

(e) A capacitor with two different layers of dielectric is equivalent to two capacitors (each having one of the layers of the dielectric) connected in series.

(f) When using multiple dielectrics in coaxial cables the layers should be arranged so that their permittivities are in ascending order from the core outwards.

(g) The lines of force between two parallel, oppositely charged wires take the form of circles.

(h) Between two equal and opposite line charges there is a plane of zero potential which lies on the perpendicular bisector of the horizontal displacement between the lines.

(i) A conductor may be placed on any equipotential surface without affecting the field in which it lies.

(j) A conducting surface may only be placed on a zero equipotential surface without affecting the field in which it lies.

(k) Multiplate capacitors are used as a means of effectively increasing plate area and so increasing capacitance.

(l) In a multiplate capacitor there are effectively as many 'capacitors' as there are plates.

(m) In a multiplate capacitor the 'capacitors' are effectively connected in parallel.

46

> True: b, d, e, g, h, i, k, m

(a) It is maximum at the surface of the inner conductor.   (Frame 29)

(c) The image conductor is of the opposite sign.   (Frame 38)

(f) It should be in *descending* order.   (Frame 42)

(j) It may be placed on any equipotential surface without affecting the field which is everywhere at right angles to it.   (Frame 38)

(l) There is one less 'capacitor' than the number of plates.   (Frame 45)

47

SUMMARY OF FRAMES 25–46

Configuration	Capacitance
Coaxial cables concentric cylinders	$\dfrac{2\pi\varepsilon_0\varepsilon_r}{\ln(b/a)}$ (F m^{-1})
Parallel wires Parallel cylinders Parallel line charges	$\dfrac{\pi\varepsilon_0\varepsilon_r}{\ln(d/a)}$ (F m^{-1})
Wire and parallel conducting plate	$\dfrac{2\pi\varepsilon_0\varepsilon_r}{\ln(d/a)}$ (F m^{-1})
Parallel plates with two layers of dielectric	$\dfrac{A\varepsilon_0\varepsilon_{r1}\varepsilon_{r2}}{\varepsilon_{r2}d_1 + \varepsilon_{r1}d_2}$ (F)
Multi-plate, having N plates arranged in 'parallel'	$\dfrac{A\varepsilon_0\varepsilon_r}{d}(N-1)$ (F)
Concentric spheres	$\dfrac{4\pi\varepsilon_0\varepsilon_r ab}{b-a}$ (F)
Isolated sphere	$4\pi a\varepsilon_0\varepsilon_r$ (F)

In a composite capacitor, the higher stress occurs in the dielectric material having the lower permittivity.

Concerning coaxial cables:

at any point in the dielectric, $E = V/(x \ln(b/a))$;
the maximum stress occurs at the surface of the inner conductor (i.e. $x = a$);
for economy of dielectric material, make $b/a = e$;
if the cable is graded, the layers of dielectric should be arranged so that the one with the highest relative permittivity is nearest the core.

In any mixed dielectric capacitor, the highest stress occurs in the layer having the lowest permittivity.

48

EXERCISES

(i) Calculate the capacitance of 1 km of a coaxial cable whose inner conductor has a diameter of 3 mm and whose outer conductor has an inside diameter of 1 cm. The relative permittivity of the dielectric material is 2.5.

(ii) The sheath of a coaxial cable has an inside diameter of 7.6 cm and is to operate at 90 kV. Determine the radius of the conductor which will produce the least electric stress at its (the conductor's) surface and calculate the value of this stress.

(iii) Calculate the capacitance per metre length of a twin line whose conductors each have a radius of 1.6 mm and are spaced 0.5 m apart in air. The effect of the earth may be neglected.

(iv) A parallel plate capacitor consists of two metal plates 10 cm square, 8 mm apart. Half of this space is taken up by a slab of dielectric having a relative permittivity of 6, the remainder being air. A potential difference of 400 V is maintained across the plates. Calculate (a) the capacitance of the capacitor and (b) the p.d. across the dielectric.

(v) A coaxial cable consists of an inner conductor of diameter 10 mm, a layer of dielectric 20 mm thick and the outer conductor. The dielectric has a relative permittivity of 4 and a maximum field strength of 2 kV mm^{-1}. Calculate the capacitance per metre of the cable and its breakdown voltage.

(vi) Calculate the capacitance between concentric spherical conductors of diameter 18 cm and 20 cm when the space between them is filled with an insulating material having a relative permittivity of 2.5.

(vii) Calculate the capacitance of each of the spheres of question (vi) when they are isolated.

49

ENERGY CONSIDERATIONS

We have seen that when a body is charged, work is done and energy is stored. This energy is recoverable when the charges return to their original positions. We shall now obtain some expressions for the energy stored in a system of charged bodies.

We consider an uncharged body to be in the process of being charged to a potential of V volts by gradually placing a charge of Q coulomb upon it. Assume that, after a certain time in the charging process, the charge on the body is xQ coulomb where x has a value between 0 and 1. At that time the potential of the body will have reached xV volt.

How much work will be required to move the next fraction of charge, $dx\,Q$ coulomb, from a point of zero potential and place it on the body? (Remember that 1 joule of work is required to move a change of 1 coulomb through a potential of 1 volt.)

50

$$\boxed{(dx)Q \times (xV) \qquad \text{(J)}}$$

This is because the energy in joule is the product of the charge being moved and the potential through which it is moved. In this case we are moving the next fraction $(dx\,Q)$ from zero to a potential of xV volt (i.e. through xV volt). The total work required to charge the body to its final potential of V volt (and its final charge of Q coulomb) is obtained by adding all of the amounts of work like that in the box as x takes all values from 0 to 1. In other words we must integrate the expression in the box with respect to x from $x = 0$ to $x = 1$.

$$\text{Work required} \qquad = \int_0^1 xV \times Q\,dx \qquad = VQ \int_0^1 x\,dx$$

$$= \frac{1}{2}[x^2]_0^1 VQ \qquad = \frac{1}{2}QV \qquad \text{(J)}$$

This, then, is the energy stored and is recoverable.

What is the energy stored in terms of the capacitance of the body?

$$\boxed{CV^2/2 \quad \text{(J)}}$$

Since $\qquad Q = CV \qquad$ then $\qquad QV/2 = (CV)V/2 = CV^2/2 \qquad$ (J)

This is the energy stored on a body having capacitance C and charged to V (V). The expression applies equally to a capacitor having a capacitance of C with a potential difference of V maintained between its plates.

For a system of n capacitors the total energy stored is given by

$$W = \frac{1}{2} \sum_{a=1}^{n} C_a V_a^2 \qquad \text{(J)}$$

Why do you think that electrostatic batteries would not be much use for motor vehicles? (*Hint*: consider a 40 A h, 12 V battery.)

> The amount of energy that can be stored in a battery of reasonable dimensions is too small.

A fully charged 40 A h, 12 V battery stores 1.728 MJ. A 12 V capacitor would have to have a capacitance of 24 kF(!) to store this amount of energy. Try designing a parallel plate capacitor with this amount of capacitance.

It is useful to have expressions for the stored energy in terms of the electric field strength and/or the electric flux density. For a parallel plate capacitor we can derive such expressions from the expression

$$W = \frac{1}{2} CV^2 \qquad \text{(J)}$$

Using the expression for the capacitance of a parallel plate capacitor and remembering that $V = Ed$, see if you can obtain an expression for the energy stored in terms of E and D.

53

$$\boxed{ED Ad/2 \qquad \text{(J)}}$$

Here's how its done:

Energy stored, $\qquad W = \dfrac{1}{2}CV^2 = \dfrac{1}{2}\dfrac{A\varepsilon}{d}(Ed)^2 \qquad \text{(J)}$

$$= \dfrac{1}{2}\dfrac{A\varepsilon}{d}(ED/\varepsilon)d^2 = \dfrac{1}{2}EDAd \qquad \text{(J)}$$

$A \times d$ is the volume of the electric field between the plates so that we can write $W = \frac{1}{2}ED \times$ the volume of the field $= \frac{1}{2}ED$ $\qquad$ (J m^{-3}).

Now obtain expressions for the energy stored in terms of (a) E and ε and (b) D and ε.

54

$$\boxed{\text{(a) } \varepsilon E^2/2; \text{ (b) } D^2/2\varepsilon \text{ (both in J m}^{-3}) \dots \text{just use } D = \varepsilon E}$$

These expressions were obtained by considering, in Frame 53, a parallel plate capacitor. To show that they hold for any arrangement of conductors we consider a 'tube' of lines of force between two conductors of any shape whatever as shown in the diagram. This Faraday tube, as it is called, has charge amounting to $+dq$ coulomb at its end on conductor A and $-dq$ coulomb at its end on conductor B. The flux in the tube is thus dq coulomb.

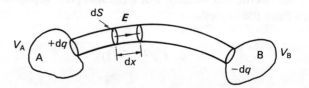

If the conductors A and B are respectively at potentials V_A and V_B, what is the energy due to the charges at the ends of the tube?

$$\boxed{dW = dq\,(V_A - V_B)/2 \qquad \text{(J)}}$$

The flux crossing any surface such as dS in the tube is given by $dq = D\,dS$ where D is the flux density at right angles to dS. Also, $V_A - V_B$ is the sum of all the products $E\,dx$ where dx is a small length of the tube and E is the electric field strength acting along it. Substituting these into the expression in the box, we get $dW = \frac{1}{2}D\,dS\,E\,dx = \frac{1}{2}DE\,dv$ (where $dv = dx \times dS$, the volume of the small length, dx, of the tube).

There will be pairs of charges like dq all around the surfaces of the conductors A and B and a tube of flux will exist between each pair. To find the total energy stored in the electric field due to the charged conductors means adding up all the energies in all the Faraday tubes which go to make up the whole space of the field between and around the conductors. This can be written as the volume integral:

$$W = \iiint_v \frac{1}{2}DE\,dv = \iiint_v \frac{1}{2}\varepsilon E^2\,dv \qquad \text{(J)}$$

We can see from this result that, for a given amount of stored energy, the volume of the field is inversely proportional to the square of the electric field strength, E.

If the electric field strength is increased by a factor n, then the volume of the field is reduced by a factor n^2 for the same amount of stored energy.

Example

A capacitor consists of two metal plates each of area 0.36 m^2 separated by a slab of dielectric 2 cm thick and having $\varepsilon_r = 3$. Calculate the energy stored in the capacitor when a p.d. of 25 kV is applied to its plates. If E is doubled, what must be the area of square plates for the same amount of stored energy and the same plate separation?

Solution

$$W = \frac{1}{2}\varepsilon E^2\,Ad = \frac{(3 \times 8.854 \times 10^{-12}) \times (625 \times 10^6) \times 0.36 \times 0.02}{2 \times (0.02)^2} = 0.149\ \text{J}.$$

If E is doubled, the volume is one-quarter of its previous value so that, for the same separation, the area will be a quarter of the previous value. The sides of the square plates will thus be 30 cm.

56

MECHANICAL FORCES IN ELECTROSTATIC FIELDS

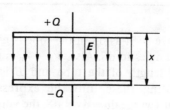

The capacitor shown has plates of area A and surface charge density σ. As they are oppositely charged, there will be a force of attraction between them. We let an external force, F, be used to increase the separation slightly by dx.

Before the separation is increased, energy stored $= \frac{1}{2}DEAx$ (J).

After the increase, energy stored $= \frac{1}{2}DEA(x + dx)$ (J).

The *increase* in stored energy $dW = \frac{1}{2}DEA\,dx = \frac{1}{2}D\frac{D}{\varepsilon}A\,dx$ (J).

Neglecting end effects, $D = \sigma$ so that $dW = \frac{1}{2}\frac{D^2}{\varepsilon}A\,dx = \frac{1}{2}\frac{\sigma^2}{\varepsilon}A\,dx$ (J).

This increase in stored energy must equal the work done by the external force in separating the plates. Write down the energy balance equation.

57

$$F\,dx = \frac{\sigma^2}{2\varepsilon}A\,dx$$

Hence $\qquad\qquad\qquad F = \dfrac{\sigma^2}{2\varepsilon} \times A\ (\text{N}) = \dfrac{D^2}{2\varepsilon} \times A\ (\text{N}).$

Also, using $D = \varepsilon E$, $\qquad F = \dfrac{1}{2}DEA\ (\text{N}) = \dfrac{1}{2}\varepsilon E^2 A\ (\text{N}).$

What is the maximum force possible in air? (Air breaks down if $E > 3$ MV m^{-1}.)

58

> 39.84 N m^{-2} (the pressure of a column of air 3.2 m long)

Example

A capacitor consists of parallel plates of area 70 cm × 55 cm separated by a material 2 mm thick and having a relative permittivity of 3. A potential difference of 5 kV is maintained between the plates. Calculate the force between them.

Solution

The force is attractive and its magnitude is given by

$$F = \frac{1}{2}\varepsilon E^2 A = \frac{1}{2}\varepsilon \frac{V^2}{d^2} A = \frac{3 \times 8.854 \times 10^{-12} \times 25 \times 10^6 \times 0.7 \times 0.55}{2 \times 4 \times 10^{-6}} = 31.95 \text{ N}$$

59

BOUNDARY CONDITIONS BETWEEN TWO DIFFERENT MEDIA IN ELECTROSTATIC FIELDS

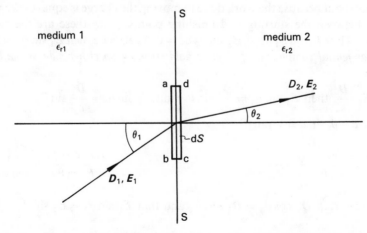

In the diagram, SS is a plane surface separating two dielectric materials of relative permittivities ε_{r1} and ε_{r2}. The cross-section of a thin Gaussian surface (whose plane area is dS) is shown as abcd and this Gaussian surface encloses a small piece of each dielectric. It encloses no charge.

What can you deduce about the relationship between D_1 and D_2?

60

$$\boxed{D_1 \cos \theta_1 = D_2 \cos \theta_2}$$

Since the charge enclosed by the Gaussian surface is zero then the net flux out of it is zero, i.e. $D_2 \cos \theta_2 \, dS - D_1 \cos \theta_1 \, dS = 0$. This means that $D_1 \cos \theta_1 = D_2 \cos \theta_2$ and *the normal component of flux density on either side of the surface is the same.*

Suppose we now move a unit positive charge along the path abcda. The work done is $E_1 \sin \theta_1 \, ab + E_1 \cos \theta_1 \, bc - E_2 \sin \theta_2 \, cd - E_2 \cos \theta_2 \, da$. Since bc = da ≈ 0, this reduces to $E_1 \sin \theta_1 \, ab - E_2 \sin \theta_2 \, cd$.

What can you deduce about the relationship between E_1 and E_2?

61

$$\boxed{E_1 \sin \theta_1 = E_2 \sin \theta_2}$$

This comes about because the work done in moving the charge is equal to the potential difference between the starting and finishing points. Since these are the same then the p.d. = 0. Thus $E_1 \sin \theta_1 \, ab - E_2 \sin \theta_2 \, cd = 0$. As ab = cd then $E_1 \sin \theta_1 = E_2 \sin \theta_2$ and the *tangential component of electric field strength on either side of the boundary is the same.*

Since $E = \dfrac{D}{\varepsilon}$, then $E_1 \sin \theta_1 = \dfrac{D_1}{\varepsilon_1} \sin \theta_1$ and $E_2 \sin \theta_2 = \dfrac{D_2}{\varepsilon_2} \sin \theta_2$.

Also since $\cos \theta / \sin \theta = 1/\tan \theta$:

$$\frac{D_1 \cos \theta_1}{E_1 \sin \theta_1} = \frac{\varepsilon_1}{\tan \theta_1} \quad \text{(i)} \qquad \text{and} \qquad \frac{D_2 \cos \theta_2}{E \sin \theta_2} = \frac{\varepsilon_2}{\tan \theta_2} \quad \text{(ii)}$$

We have seen that $D_1 \cos \theta_1 = D_2 \cos \theta_2$ and that $E_1 \sin \theta_1 = E_2 \sin \theta_2$

so that $$\frac{D_1 \cos \theta_1}{E_1 \sin \theta_1} = \frac{D_2 \cos \theta_2}{E_2 \sin \theta_2} = \frac{\varepsilon_2}{\tan \theta_2} \quad \text{(iii)}$$

Obtain the relationship between $\varepsilon_1, \varepsilon_2, \theta_1$ and θ_2.

$$\boxed{\frac{\varepsilon_1}{\varepsilon_2} = \frac{\tan \theta_1}{\tan \theta_2}}$$

This is obtained from (i) and (iii) in the previous frame and is known as the *law of electric flux refraction* at the boundary or surface of discontinuity.

Example

Two parallel sheets of glass have a uniform air gap between their inner surfaces and are sealed around their edges. They are immersed in oil and mounted vertically. The relative permittivities of the oil and the glass are respectively 2.5 and 6. Calculate the *E*-vectors in the glass and in the air when that in the oil is 1 kV m^{-1} and enters the glass at 60° to the horizontal.

Solution

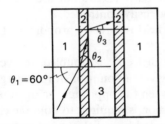

medium 1 is oil
medium 2 is glass
medium 3 is air

Using the law of electric flux refraction, we have:

$$\tan \theta_2 = (\varepsilon_{r2}/\varepsilon_{r1}) \tan \theta_1 = (6/2.5) \tan 60° = 4.16 \text{ and } \theta_2 = 76.5°$$
$$\tan \theta_3 = (\varepsilon_{r3}/\varepsilon_{r2}) \tan \theta_2 = (1/6) \tan 76.5° = 0.693 \text{ and } \theta_3 = 34.8°$$

Using the relationship in the box at the top of Frame 60 together with the relationship between *D* and *E* we get:

$$\boldsymbol{D}_1 \cos \theta_1 = \varepsilon_0 \varepsilon_{r1} \boldsymbol{E}_1 \cos \theta_1 = \varepsilon_0 \varepsilon_{r1} \, 1000 \times 0.5 = \varepsilon_0 \varepsilon_{r1} \, 500 \text{ C m}^{-2}$$

and $$\boldsymbol{D}_1 \cos \theta_1 = \boldsymbol{D}_2 \cos \theta_2 = \varepsilon_0 \varepsilon_{r2} \boldsymbol{E}_2 \cos \theta_2 = \varepsilon_0 \varepsilon_{r1} \, 500 \text{ C m}^{-2}$$

$$\therefore \boldsymbol{E}_2 = \frac{500 \, \varepsilon_{r1}}{\cos \theta_2 \, \varepsilon_{r2}} = \frac{500 \times 2.5}{0.2334 \times 6} = 892 \text{ V m}^{-1}$$

Similarly, $$\varepsilon_0 \varepsilon_{r3} \boldsymbol{E}_3 \cos \theta_3 = \varepsilon_0 \varepsilon_{r2} \boldsymbol{E}_2 \cos \theta_2$$

whence $$\boldsymbol{E}_3 = 1523 \text{ V m}^{-1}.$$

63

Which of the following statements are true?

(a) When a charge of 1 coulomb is moved through a potential difference of 1 volt the energy required is 1 joule.

(b) The energy stored in a capacitor of C farad having a charge of Q coulomb and with a potential difference of V volt between its plates is $QV/2$ (J).

(c) The energy stored in a capacitor having a capacitance of C farad, charged to V volt is given by $W = (CV/2)$ multiplied by the volume of the field between the plates and is measured in joule.

(d) The energy stored in an electric field of density D coulomb per square metre and relative permittivity ε_r is given by

$$W = \frac{D^2}{2\varepsilon_0 \varepsilon_r} \text{ joule per cubic metre of the field}$$

(e) The only reason why electrostatic batteries are not used for motor cars is that the rate of energy recovery is not easily controlled.

(f) The energy stored in an electric field is directly proportional to the volume of the field.

(g) If the electric field strength in a dielectric is doubled, the volume of the field need be only half as great for a given amount of stored energy.

(h) There is a force of repulsion between the oppositely charged plates of a parallel plate capacitor so that encapsulation is required to hold the capacitor together.

(i) If a pair of parallel plates is separated by a dielectric of permittivity ε and if the electric field strength in the dielectric is E, then there will be a force between the plates given by

$$F = \frac{1}{2}\varepsilon_0 \varepsilon_r E^2 \quad (\text{N m}^{-2})$$

(j) The force exerted by an electric field is independent of the flux density so long as the medium of the field is a vacuum or air.

(k) Mechanical forces in electric fields are relatively weak.

(l) At a boundary between two different media in an electric field, the tangential component of flux density on either side of the boundary is the same.

(m) At a surface of discontinuity in an electric field, the lines of force are more nearly perpendicular to the surface on the side where the permittivity is higher.

(n) At a boundary between two different media in an electric field, the normal component of electric field strength on either side of the boundary is the same.

(o) Lines of force are refracted at a surface of discontinuity in an electric field.

64

True: a, b, d, f, i, k, o

(c) $W = (CV^2)/2$ (J).   (Frame 51)
(e) The amount of energy which can be stored in a capacitor of
      reasonable size is too small.   (Frame 52)
(g) The volume of the field $\propto 1/E^2$.   (Frame 55)
(h) The force is one of attraction.   (Frame 56)
(j) The force is proportional to the flux density squared.   (Frame 57)
(l) The normal component of D is the same on either side.   (Frame 60)
(m) They are more nearly perpendicular to the surface on the
      side of lower permittivity.   (Frame 62)
(n) The tangential component on either side is the same.   (Frame 61)

65

SUMMARY OF FRAMES 49–64

Energy stored in an electric field

$$W = CV^2/2 = QV/2 = Q^2/2C \quad \text{(J)}$$

$$W = ED/2 = \varepsilon_0 \varepsilon_r E^2/2 = D^2/(2\varepsilon_0 \varepsilon_r) \quad \text{(J m}^{-3} \text{ of the field)}$$

Force in the electric field

$$F = ED/2 = \varepsilon_0 \varepsilon_r E^2/2 = D^2/(2\varepsilon_0 \varepsilon_r) \quad \text{(N m}^{-2})$$

At a boundary between different media in an electric field

The normal component of electric flux density is the same on either side of the boundary or surface of discontinuity.

The tangential component of electric field strength is the same on either side of the boundary or surface of discontinuity.

The law of electric flux refraction:

$$\frac{\tan \theta_1}{\tan \theta_2} = \frac{\varepsilon_1}{\varepsilon_2}.$$

66

EXERCISES

(i) Calculate the energy required to charge a capacitor of 50 pF to a potential difference of 600 V.

(ii) A capacitor consists of two flat metal plates each 20 cm × 20 cm in area separated by an air space of 2 mm thickness. Determine the energy stored in the capacitor when a potential difference of 400 V is applied between the plates.

(iii) Two metal plates each of 35 cm × 25 cm are placed 10 mm apart in air and a potential difference of 4000 V is maintained between them. Determine the force of attraction between the plates.

(iv) Two plane surfaces in an electric field are separated by three layers of dielectric, A, B and C having relative permittivities of 2, 3 and 4 respectively. Calculate the magnitude and direction of the electric field strength in the dielectrics B and C when the electric field strength in dielectric A is 5 kV m^{-1} directed towards B:

(a) at 30° to the horizontal at the A/B boundary,

(b) at right angles to the A/B boundary.

67

SHORT EXERCISES ON PROGRAMME 3

The number on the right gives the frame where the answer may be found.

1. Explain why metals are good conductors of electricity. (2)
2. Why can there be no component of electric field strength tangential to the surface of a conductor in an electrostatic field? (2)
3. State with reasons whether it is possible for charge to be uniformly distributed within the volume of a conductor in an electrostatic field. (3)
4. What is the unit of surface charge density? (4)
5. What is the relationship between electric flux density and surface charge density immediately outside the surface of a conductor? (5)
6. Explain what is meant by 'capacitance'. (6)
7. Give the relationship between capacitance, charge and voltage. (6)
8. Define a capacitor. (7)
9. What are the factors governing the capacitance of a capacitor? (8)
10. What is the unit of capacitance? (7)
11. What is meant by a 'dielectric'? (8)

12. Give the relationship between ε, ε_0 and ε_r. (10)
13. What is the unit of ε_0? (14)
14. What is the effect of doubling the plate area of a capacitor? (15)
15. What is meant by 'fringing' in parallel plate capacitors? (14)
16. If the plate separation of a parallel plate capacitor is reduced, what happens to the capacitance? (15)
17. What is the effective capacitance of five identical capacitors each of 50 pF when they are connected in parallel? (19)
18. Why are multiplate capacitors used? (45)
19. What is the effect of using composite (mixed dielectric) capacitors? (41)
20. Where in a coaxial cable is the electric field strength a maximum? (29)
21. What is the condition for the maximum electric field strength in the dielectric of a coaxial cable to be as small as possible? (30)
22. What is meant by an electric 'image'? (38)
23. Give an expression for the energy stored in a capacitor in terms of its capacitance and the potential difference between its plates. (51)
24. Why are electrostatic batteries unsuitable for motor vehicles? (52)
25. Give an approximate value for the maximum force in an electrostatic field in air, given that air breaks down at 3 MV m^{-1}. (58)
26. What is meant by electric flux refraction? (62)
27. State the relationship between the electric field strength on either side of a boundary between different media in an electric field. (61)
28. State the relationship between the electric flux density on either side of a surface of discontinuity in an electric field. (60)

68

ANSWERS TO EXERCISES

Frame 24

(i) 1.859×10^{-11} F m^{-1}; (ii) 4.52; (iii) 6×10^{-8} C; (iv) 1.67 V; (v) 3.8 pF; (vi) 0.89 μC; (vii) 354 pF; (viii) (a) 3.54 nF, (b) 1.77 μC, (c) 177 μC m^{-2}, (d) 5000 kV m^{-1}; (ix) 3, 9, 27, 12, 30, 36, 2.25, 2.7, 6.75, 29, 2.08, 2.77, 6.9, 8.31 (all answers in μF).
39

Frame 48

(i) 115.5 nF; (ii) 1.398 cm, 64.37 kV cm^{-1}; (iii) 4.84 pF m^{-1}; (iv) 18.97 pF, 57.14 V; (v) 138 pF, 16.1 kV; (vi) 250 pF; (vii) 11.13 pF; 10.01 pF.

Frame 66

(i) 9 μJ; (ii) 14.17 μJ; (iii) 61.98 mN; (iv) (a) 3.82 kV m^{-1} at 40.9° to the horizontal, 3.297 kV m^{-1} at 49° to the horizontal, (b) 3.33 kV m^{-1}, 2.5 kV m^{-1} (both in the same direction as the field strength in A).

Programme 4

CONDUCTION FIELDS

1

INTRODUCTION

Programmes 2 and 3 dealt with charges at rest. We now consider the effects of charges being in motion. Charges in motion give rise to what we call electric currents and these can be steady currents or currents which vary with time. In this programme we are concerned with *steady* currents by which we mean constant, direct currents. Time varying currents can have *steady STATE* values which means that the time variations are repeated periodically. Take care not to confuse the two. The fields associated with steady currents are static (i.e. constant with time).

The following topics will be covered in this programme:

- electric current and current density
- the law of conservation of charge (the continuity equation)
- conduction field lines of force and tubes of flow
- Ohm's law
- resistance and resistivity
- conductance and conductivity
- resistances in series and parallel
- the Joule effect
- temperature coefficient of resistance
- Kirchhoff's circuit laws
- the analogies between the electrostatic and the electroconductive fields.

When you have studied this programme you should be able to:

- distinguish between convection and conduction currents
- calculate the current density in a conductor
- describe what is meant by an ohmic or linear material
- calculate the resistance of a piece of material given its physical dimensions and its resistivity
- calculate the leakage resistance of a parallel plate capacitor
- calculate the leakage resistance per metre length of a coaxial cable
- obtain from first principles an expression for the resistance of the material between any system of conductors (e.g. concentric spheres)
- calculate the equivalent resistance of a number of resistors connected in series, parallel or a combination of these
- calculate the power dissipated in any system of resistors
- calculate the resistance of a resistor at any temperature, given the resistance at some other temperature and the temperature coefficient of resistance
- state Kirchhoff's circuit laws and apply them to simple d.c. circuits
- use the method of field analogies to convert expressions for capacitance into expressions for conductance for given conductor geometries.

2

ELECTRIC CURRENT

When electric charges are in motion they constitute an electric current. A steady flow of electric charges which does not change periodically with time is called a direct current. The symbol for current is I and its SI unit is the ampere (A). One ampere is the rate of movement of charge, amounting to one coulomb per second passing a given plane of reference.

Give a mathematical relationship between electric current, I, and electric charge, Q.

3

$$\boxed{I = \frac{dQ}{dt}}$$

It follows from this relationship that the charge which is moved when a current of I ampere flows for T seconds is given by

$$Q = \int_0^T I \, dt \qquad \text{over the time } T$$

so that $Q = I \times T(C)$.

There are two types of electric current. One is called convection current and is due to the movement of charged particles through space. It does not, therefore, take place in metals but occurs in, for example, electron beams in cathode ray oscilloscopes. The other type is called conduction current and is due to the drift of charges through the atomic lattice sub-structure of a medium such as copper.

The rest of this programme deals exclusively with conduction currents and when we speak of current it is to be understood to mean conduction current. The conventional positive direction of current flow is the direction of the electric field strength (the E-vector direction). In metals the electric current is due to the movement of electrons through them. Is the direction of electric current flow the same as the direction of electron drift?

4

> No. The direction of the *E*-vector is the direction in which a unit positive charge would move. Electrons are negative charges and therefore move in the opposite direction.

CURRENT DENSITY

The current per unit area passing at right angles to a given plane of reference is called the current density (symbol, J; unit, ampere per square metre $(A\ m^{-2})$)) which is a *vector* quantity.

ELECTRON DRIFT VELOCITY

When an electric current is switched 'on' the effect is virtually instantaneous so it may be thought that electrons move very fast through the material (copper say). Let us see if this is in fact the case by determining the drift velocity of the electrons in a piece of copper wire. We shall assume a current density of $1.3 \times 10^6\ A/m^2$ (a commonly used value). In copper there is about $1.3 \times 10^{10}\ C$ of charge per cubic metre due to the number of electrons free to move.

Now the unit of velocity is the metre per second and we will manipulate this a little so as to introduce the information given, i.e. current density $(A\ m^{-2})$ and charge volume density $(C\ m^{-3})$:

velocity,
$$v = \frac{m}{s} = \frac{m}{s} \times \frac{A\ m^{-3}}{A\ m^{-3}} = \frac{A\ m^{-2}}{A\ s\ m^{-3}}$$

But ampere second $(A\ s) = $ Coulomb (C):

$$\therefore \quad \frac{m}{s} = \frac{A\ m^{-2}}{C\ m^{-3}}$$

So what is the drift velocity of the electrons in the piece of wire under consideration?

5

$$\boxed{\text{Approximately } 10^{-4} \text{ m s}^{-1}}$$

$$v = \frac{\text{current density}}{\text{charge volume density}} = \frac{1.3 \times 10^6}{1.3 \times 10^{10}} = \underline{10^{-4} \text{ m s}^{-1}}$$

So the electrons most certainly do not move very fast!

This velocity (about one-tenth of a millimetre per second) is the average drift velocity of the electrons through the conductor. It is *not* the velocity of all of the electrons or of any particular electron.

For an electric current to flow along a conductor, therefore, it is not necessary for there to be motion of an electron from one end of it to the other but that the forces on the electrons are transmitted from one to another.

6

THE LAW OF CONSERVATION OF CHARGE

The closed surface S shown in the diagram is made up of an infinite number of small elements of surface such as dS.

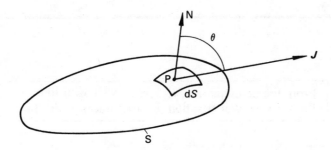

If the surface encloses charges which are in motion then the current, I, crossing the surface element dS at right angles to it is given by $J \cos \theta$ dS. J is the current density at P, a point on dS, and is measured in ampere per metre squared (A m^{-2}). $J \cos \theta$ is the component of J in the direction of PN, the normal to the surface dS at P.

To calculate the total current (I_S, say) flowing out of the whole surface S, we multiply every element like dS of which it consists by the component of current density at right angles to the surface there and add the results. Express this as a surface integral.

7

$$I_S = \int_S J \cos \theta \, dS$$

The current flowing out of the surface S is a rate of flow of charge and must therefore equal the rate of decrease of charge inside the surface. If the volume enclosed by the surface S is v and the charge per unit volume is ρ then the charge enclosed by S is ρv and the rate of decrease of this charge is

$$-\frac{\partial(\rho v)}{\partial t} \quad \text{so that } I_S = -\frac{\partial(\rho v)}{\partial t}$$

The partial derivative is used because ρ may vary not only with time but also with its position.

We have, then, that

$$I_S = \int_S J \cos \theta \, dS \quad \text{and that } I_S = -\frac{\partial(\rho v)}{\partial t}.$$

Bearing in mind that we are dealing with steady current flow so that J is not varying with time, what conclusions can you draw from these two equations?

8

From the first equation, I_S doesn't vary with time.
From the second equation, ρ varies linearly with time.

The first conclusion is straightforward. For the second we realise that if I_S doesn't vary with time then $-\dfrac{\partial(\rho v)}{\partial t}$ is constant and so is $-\dfrac{\partial(\rho)}{\partial t}$ because v is a constant. It follows that ρ must vary linearly with time. However, this cannot be so if there is *steady* current flow which means steady charge flow. Thus $-\dfrac{\partial(\rho v)}{\partial t} = 0$ and therefore $I_S = 0$. What, then, is the integral of J over any closed surface?

9

$$\int_S J \cos \theta \, dS = 0$$

This is the *continuity equation* and is one of the two basic equations of conduction field theory. The other one is a special case of the formula for the electric potential difference between two points A and B which we derived in Programme 2, Frame 37. Where the two points are coincident the formula is written $\oint_L E \cos \theta \, dl = 0$ and is read 'the line integral of E around any *closed* path L is equal to zero'. This is called the circuit law.

10

Conduction fields can be pictured using 'lines of force' in the same way as for electrostatic fields. The direction of a line of force at a point in a conduction field is the direction of the E-vector at that point and, just as is the case in electrostatic fields, they begin and end on charges. They are always at right angles to equipotential surfaces.

All of the lines of force between two areas in a field can be thought of as forming a 'tube' of flux. In the diagram, the two areas dS_1 and dS_2 together with the outer lines of force of the tube form a closed surface, S.

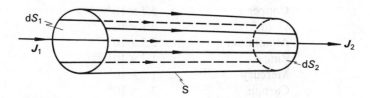

J_1 is the current density at right angles to the surface dS_1.
J_2 is the current density at right angles to the surface dS_2.
Applying the continuity equation to the surface S:

$$J_1 \, dS_1 - J_2 \, dS_2 = 0$$

$$\therefore \quad J_1 \, dS_1 = J_2 \, dS_2$$

This means that the current crossing surface dS_1 equals the current crossing surface dS_2. Since these were arbitrarily chosen surfaces, it follows that the current is the same everywhere in the tube.

11

OHM'S LAW

Experiment shows that for many conductor materials the current density J is proportional to the electric field strength E, i.e. $J \propto E$. These materials are known as *linear* or *ohmic* materials.

We write $J = \sigma E$ where σ is the constant of proportionality and is called the conductivity of the material. Its SI unit is the siemens per metre (S m^{-1}). The relationship $J \propto E$ is one form of Ohm's Law.

The reciprocal of conductivity is resistivity, symbol ρ and the reciprocal of the siemens is the ohm. What is the unit of resistivity?

12

The ohm-metre

There is an enormous range of values of resistivity (and therefore of conductivity) as the following examples will illustrate. For a given material, the values vary with temperature, the values given here being for 20°C.

Material	Conductivity (S m^{-1})
Silver	6.1×10^7
Copper	5.7×10^7
Gold	4.1×10^7
Aluminium	3.5×10^7
Zinc	1.7×10^7
Mercury	1×10^6
Carbon	3×10^4
Sea water	4
Distilled water	1×10^{-4}
Bakelite	1×10^{-9}
Glass	1×10^{-12}
Mica	1×10^{-15}
Quartz	1×10^{-17}

Good conductors have a high value of conductivity whereas good insulators have a low conductivity.

What are the values of the resistivities at 20°C of (i) the best conductor and (ii) the best insulator in the above table?

(i) 1.64×10^{-8} Ω-m (silver); (ii) 10^{17} Ω-m (quartz)

The best conductor in the table is the one with the highest conductivity (silver) and its resistivity is the reciprocal of 6.1×10^7 S m^{-1}. Similarly, the resistivity of the best insulator in the table (quartz) is the reciprocal of 1×10^{-17} S m^{-1}.

ANOTHER FORM OF OHM'S LAW

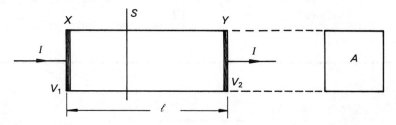

The diagram shows a conductor of length l and cross-sectional area A. Current is fed into the left-hand side and out of the right-hand side through plates X and Y, each of which have infinite conductivity. Plate X is an equipotential surface of potential V_1 and plate Y is a surface of potential V_2. S is an arbitrarily chosen cross-section of the conductor.

The current everywhere in the 'tube of flow' is the same (see Frame 10) and is given by $I = \int_S J \, dS$ where J is the current density at right angles to dS, an element of the cross-section.

Since $J = \sigma E$ then

$$I = \int_S \sigma E \, dS \qquad \text{(i)}$$

The potential difference between the two plates X and Y is given by the circuit law as

$$V_1 - V_2 = \int_L E \, dl \qquad \text{(ii)}$$

where E is the component of electric field strength in the direction of dl, an element of the path L (any path between X and Y).

What do the two equations, (i) and (ii), tell us about the relationship between $V_1 - V_2$ (call it V) and I?

14

$$\boxed{V \propto I}$$

Equation (i) shows that I is a linear function of E and equation (ii) shows that V is a linear function of E. Any change in I therefore results in a proportionate change in E and, in turn, a proportionate change in V so that $V \propto I$, which is another form of Ohm's Law.

We write $V = RI$ where R is a constant of proportionality and is called the resistance of the conductor. Its SI unit is the ohm (Ω).

The reciprocal of resistance is conductance (G) whose SI unit is the siemens (S). What is the relationship between G, V and I?

15

$$\boxed{G = I/V \ (\text{S})}$$

We will now consider how the resistance of a conductor depends upon its physical parameters. For the particular conductor shown in the diagram in Frame 13 we have

$$I = \int_S \sigma E \, dS \qquad \text{and} \qquad V = \int_L E \, dS.$$

Since E is uniform over the cross-section and along the length of the conductor and because σ is a constant we may write these as

$$I = \sigma E \int_S dS \qquad \text{and} \qquad V = E \int_L dl$$

Therefore

$$I = \sigma E A \qquad \text{since} \int_S dS = A$$

and

$$V = El \qquad \text{since} \int_L dl = l.$$

Now obtain an expression for the resistance, R, of the conductor in terms of its physical parameters (its length, cross-sectional area and resistivity).

$$R = \rho l / A$$

This is obtained by using Ohm's Law in the form $R = V/I$ and substituting for $V(=El)$ and $I(=\sigma EA)$, and then putting $\sigma = 1/\rho$.

It follows that the conductance, $G = 1/R = A/\rho l = \sigma A/l$ (S).

Example

A metal rod, 10 cm long and 0.75 cm in diameter, has a resistance of 95 $\mu\Omega$ at a certain temperature. If the rod is drawn out into a wire of uniform diameter 0.02 cm, calculate its resistance per metre at the same temperature.

Solution

To find the resistance per metre of the wire, we need to know the resistivity of the metal and we can find this because we know the resistance of the rod.

For the rod, then, the resistance,

$$R = \frac{\rho l}{A} \ (\Omega)$$

so that

$$\rho = \frac{RA}{l} \ (\Omega\text{-m}) = \frac{95 \times 10^{-6} \times \pi(75 \times 10^{-2})^2/4}{0.1} \ \Omega\text{-m}$$

$$= 4.2 \times 10^{-4} \ \Omega\text{-m}$$

For the wire, considering 1 m length:

$$R = \frac{\rho l}{A} \ (\Omega) = \frac{4.2 \times 10^{-4} \times 1}{\pi(2 \times 10^{-2})^2/4}$$

$$= \underline{1.33 \ \Omega}$$

If a bar measuring 6 cm $\times$ 9 cm in cross-section and 1.5 m long is made from the same metal, what are the three values of resistance between opposite faces of the bar?

17

$$\boxed{0.12 \ \Omega, \ 0.19 \ \text{m}\Omega, \ 0.42 \ \text{m}\Omega}$$

We have calculated these values by straightforward application of the formula. This is satisfactory where the cross-sectional area is uniform throughout the length of the conductor between its ends. In general, the resistance may be obtained from first principles by determining the potential difference between its ends and then dividing this by the current flowing in it.

Example

Calculate the resistance between two parallel conducting plates each measuring 20 cm × 30 cm if the space between them is filled with a slab of conducting material of thickness 3 mm and conductivity $\sigma = 5.7 \times 10^7 \ \text{S m}^{-1}$.

Solution

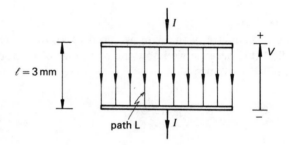

Assuming the conduction field (the flow of current) between the plates to be uniform and neglecting edge effects, we can say that the current density at all points within the conducting material is given by

$$J = \frac{I}{A} \ (\text{A m}^{-2})$$

where A is the area of each plate and I is the current flowing between them. At any point in the field distant x from one of the plates we have that

$$J_x = \sigma E_x \qquad \text{so that } E_x = \frac{J_x}{\sigma} = \frac{I}{\sigma A} \ (\text{V m}^{-1})$$

The potential difference between the plates is given by

$$V = \int_L E_x \, \mathrm{d}x = \frac{I}{\sigma A} \int_L \mathrm{d}x = \frac{Il}{\sigma A} \ (\text{V})$$

The resistance,

$$R = \frac{V}{I} = \frac{Il}{\sigma AI} = \frac{l}{\sigma A} = \frac{3 \times 10^{-3}}{0.2 \times 0.3 \times 5.7 \times 10^7} \, (\Omega)$$

$$= \underline{8.8 \times 10^{-10} \, \Omega}$$

Note that this could have been obtained simply by 'filling in' the formula for resistance,

$$R = \frac{\rho l}{A} \left(= \frac{l}{\sigma A} \right)$$

because the field between the plates is uniform.

Give an expression for the conductance of the material between the plates in terms of its cross-sectional area, thickness and conductivity.

18

$$\boxed{G = A\sigma/l \, (\text{S})}$$

We will now work through an example in which the field is not uniform between the conductors. Ideally, the dielectric material between the core and sheath of a coaxial cable will be a perfect insulator i.e. it will have zero conductivity. In practice this is not so and a small leakage current will flow along the **E**-field lines of force radially between the core and the sheath. Let us calculate the leakage resistance per metre length of a coaxial cable having a core radius of 1.5 mm, a sheath radius of 10 mm (and negligible radial thickness) and a dielectric material of conductivity $10^{-15} \, \text{S m}^{-1}$.

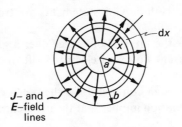

$J-$ and
$E-$field
lines

Let the leakage current be I ampere per metre length of the cable. At radius x, the current density will be

$$J_x = \frac{I}{2\pi x \times 1} \, (\text{A m}^{-2})$$

Give an expression for the electric field strength at x.

19

$$\boxed{E_x = J_x/\sigma = I/2\pi x\sigma \ (\text{V m}^{-1})}$$

The potential difference,

$$V = \int_a^b E_x \, dx \quad = \int_a^b \frac{I}{2\pi x\sigma} \, dx \ (\text{V})$$

$$= \frac{I}{2\pi\sigma} [\ln x]_a^b = \frac{I}{2\pi\sigma} \ln(b/a) \ (\text{V})$$

the resistance,

$$R = \frac{V}{I} = \frac{\ln(b/a)}{2\pi\sigma} (\Omega)$$

Putting $b = 10$ mm, $a = 1.5$ mm and $\sigma = 10^{-15}$ S/m, we get $R = 3 \times 10^{14} \ \Omega \ \text{m}^{-1}$.

Give an expression for the conductance per metre length of the dielectric material in terms of b, a and σ.

20

$$\boxed{G = \frac{1}{R} = \frac{2\pi\sigma}{\ln(b/a)} \ (\text{S m}^{-1})}$$

Now decide which of the following statements are true.

(a) Electric current is the integral of charge over a given time.
(b) Conventional current flow and electron drift in a metal take place in opposite directions.
(c) Electrons flow through conductors at the speed of light.
(d) The continuity equation implies that the net creation of electric charge in a region of space is zero.
(e) For linear conductors $V \propto I$.
(f) All solid conductors obey Ohm's Law.
(g) Ohm's Law may be stated as $J = \sigma E$.
(h) The unit of conductivity is the S m^{-1}.
(i) Conductivity is the reciprocal of resistance.
(j) The conductance of a material is inversely proportional to its length.

21

True: b, d, e, g, h, j

(a) Charge is the integral of current.   (Frame 3)
(c) They move at a snail's pace!   (Frame 5)
(f) Some do not (an example is the filament of a light bulb).
(i) Conductivity is the reciprocal of resistivity.   (Frame 11)

22

SUMMARY OF FRAMES 1–21

New quantities and units

Quantity	Symbol	Unit	Abbreviation
Current	I	Ampere	A
Current density	J	Ampere per square metre	$A\,m^{-2}$
Conductivity	σ	Siemens per metre	$S\,m^{-1}$
Resistivity	ρ	Ohm-metre	Ω-m
Resistance	R	Ohm	Ω
Conductance	G	Siemens	S

Formulae

$I = dQ/dt$

$$\oint_S J \cos\theta\, dS = 0 \text{ (the continuity equation)}$$

$$\oint_L E \cos\theta\, dl = 0 \text{ (the circuit law)}$$

$J = \sigma E; \quad V = IR$ (two forms of Ohm's Law)

$R = \rho l/A\ (\Omega); \quad G = A/\rho l\ (S)$

$R = 1/G; \quad \sigma = 1/\rho$

23

EXERCISES

(i)   How much current is represented by a charge of 100 nC passing a given plane of reference in 10 μs?

(ii) Find the current density in a uniform circular tube of current enclosing 150 mA if the diameter of the tube is 0.5 cm.

(iii) Determine the conductivity of a conductor in which the current density is 10 A cm^{-2} and in which a length of 150 m has a potential of 0.3 V between its ends.

(iv) Calculate the reistance at 20°C of 100 m of copper wire of diameter 2.59 mm. The resistivity of the copper at 20°C is 1.75×10^{-8} Ω-m.

(v)   A parallel plate capacitor has plates of area 14 cm × 10 cm separated by a distance of 3 mm. The space between the plates is filled with glass for which the conductivity is 2×10^{-12} S m^{-1}. Determine the leakage resistance of this capacitor.

(vi) Calculate the leakage resistance per metre length of a coaxial cable for which the radius of the core is 1.5 mm and the inner radius of the sheath (which may be considered to be of negligible thickness) is 10 mm. The dielectric material has a conductivity of 10^{-12} S m^{-1}.

(vii) Obtain an expression for the resistance between two concentric spheres of radii a and b $(b > a)$. The material between the spheres has a conductivity σ.

If the space between the spheres is filled with oil having a resistivity of 10^{14} Ω-m, and if $a = 12$ mm and $b = 20$ mm, calculate the resistance between the spheres.

(viii) A cube of conducting material has sides of length 1 m. Two perfectly conducting plates are attached to opposite sides of the cube. Calculate the resistance between the conducting plates if the conducting material has a conductivity of 1.1×10^7 S m^{-1}.

(ix) A bar of metal is 2 m long and has a rectangular cross-sectional area 20 cm × 10 cm. Its conductivity is 5.7×10^7 S m^{-1}. Calculate the conductance between pairs of opposite faces of the bar.

24

RESISTORS IN SERIES

A circuit element having resistance is called a resistor.

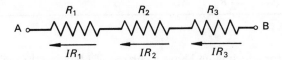

The diagram shows three resistors connected in series. The same current, I, flows through all three and the potential difference between ends A and B is given by $V = IR_1 + IR_2 + IR_3 = I(R_1 + R_2 + R_3)$.

A single resistor which would carry the same current I, and have the same potential difference between its ends, would have to have a resistance

$$R_T = (R_1 + R_2 + R_3)$$

Make up a general rule for the total or equivalent resistance of a number, say n, resistors connected in series.

25

$$R_T = \sum_{a=1}^{n} R_a \qquad \text{i.e. } R_T = R_1 + R_2 + \ldots + R_n$$

RESISTORS IN PARALLEL

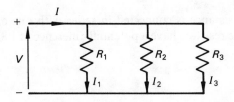

In this case, the total current, I, is the sum of the currents flowing in the individual resistors whereas the potential difference between the ends of each resistor is the same:

$$I = I_1 + I_2 + I_3 = V/R_1 + V/R_2 + V/R_3$$

Make up a rule for the total resistance of n resistors connected in parallel.

26

$$\frac{1}{R_T} = \sum_{a=1}^{n} \frac{1}{R_a} \qquad \text{i.e. } 1/R_T = 1/R_1 + 1/R_2 + \ldots + 1/R_n$$

The single resistor which would have a current of I flowing through it and a potential difference of V between its ends would need to have a resistance, R_T, such that $I = V/R_T = V/R_1 + V/R_2 + V/R_3$. Dividing throughout by V gives the answer in the box.

THE JOULE EFFECT

When a current flows through a conductor, the kinetic energy gained by the moving charges is given by the product of the charge (Q, say) and the potential difference through which it moves (V, say). If the conductor has a resistance of R ohm then we can write:

$$\text{kinetic energy gained } (KE) = Q \times V = (It) \times (IR) = I^2 Rt \text{ (J)}$$

This energy is converted into heat as the temperature of the conducting medium is raised. The rate of energy dissipation is $(KE)/t = I^2 R$ (J s^{-1}). Since $I = V/R$ then $I^2 R = V^2/R$ (J s^{-1}).

The joule per second is the watt (W) which is the unit of power and is a rate of doing work. This heating of conductors when a current flows through them is known as the Joule effect.

Example

A resistor of 20 Ω resistance is connected in series with one of 10 Ω. Calculate the power dissipated if the resistors have a potential difference of 100 V maintained across them.

Solution

Total resistance	$R = 20 \ \Omega + 10 \ \Omega = 30 \ \Omega$
Current through the combination	$I = V/R = 100/30 = 3.33$ A.
Power dissipated	$P = I^2 R = (3.33)^2 \times 30 = \underline{333 \text{ W}}$.

Repeat the problem but with the resistors connected in parallel.

$$\boxed{1500 \text{ W}}$$

Here is the working:

The equivalent resistance of the parallel combination is obtained from
$1/R_T = 1/R_1 + 1/R_2 = 1/20 + 1/10 = 3/20$.

$\therefore \quad R_T = 6.67 \ \Omega$.

$I = V/R_T = 100/6.67 = 15$ A.

The power dissipated, $P = I^2 R_T = 15^2 \times 6.67 = 1500$ W.

Alternatively, $P = V^2/R_T = 100^2/6.67 = 1500$ W.

TEMPERATURE COEFFICIENT OF RESISTANCE

It was mentioned in Frame 12 that resistivity (and therefore resistance) varies with temperature. For metals the resistance increases with temperature, for insulators it decreases with temperature, and there are some materials (for example Manganin, an alloy containing copper, manganese and nickel) for which there is practically no change in resistance over a wide range of temperature.
   For a particular material

$$R = R_S[1 + \alpha_S(T - T_S)] \ (\Omega)$$

where R is the resistance at temperature T
        R_S is the resistance at temperature T_S
        α_S is the *temperature coefficient of resistance* corresponding to T_S

and $\alpha_S = \dfrac{\text{change of resistance per degree change of temperature}}{\text{resistance at some standard temperature } (T_S)} \ (\text{degree})^{-1}$

As an example, the resistance temperature coefficient of copper is 0.0043 per degree centigrade (the standard temperature being $T_S = 0°C$). The corresponding value for manganin is 0.000 003 per degree centigrade.
   If the standard temperature is taken to be 0°C and a certain material has a resistance of R_0 at that temperature and a resistance temperature coefficient of α_0, then at temperatures T_1 and T_2 respectively its resistance will be $R_1 = R_0(1 + \alpha_0 T_1)$ and $R_2 = R_0(1 + \alpha_0 T_2)$.

It follows that $\dfrac{R_1}{R_2} = \dfrac{1 + \alpha_0 T_1}{1 + \alpha_0 T_2}$

28

KIRCHHOFF'S CIRCUIT LAWS

(a) The first (current) law

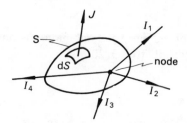

The diagram shows a number of conductors meeting at a point (such a point is called a node). S is a closed surface completely surrounding the node. Applying the continuity equation to S we have

$$\int_S J\, dS = 0$$

where J is the component of current density at right angles to dS (an element of the surface S). The left-hand side of this equation represents the total current flowing out of the surface S. Also, the total current flowing out of the node is $I_1 + I_2 + I_3 + I_4$. It follows that $I_1 + I_2 + I_3 + I_4 = 0$.

If, say, I_2 were reversed in direction, its sign would change from $+$ to $-$ so that the *algebraic sum of the currents at a node is zero*. This is Kirchhoff's current law.

(b) The second (voltage) law

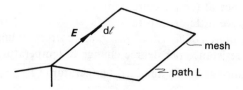

The diagram shows a closed loop of conductors forming a path L (such a loop is called a mesh). If we apply the circuit law to the path L we obtain

$$\oint_L E\, dl = 0$$

where E is the electric field strength acting along dl (an element of the path L)

Now by definition, the line integral of E along the path L is the total potential difference between its ends. Combining these two statements we arrive at Kirchhoff's voltage law. Can you deduce a statement of the law?

29

> The sum of the potential differences (voltage drops)
> around any closed loop (mesh) is zero

In general, these potential differences will consist of potential rises as well as potential drops. The currents flowing through resistance produce potential *drops* and energy is dissipated. This energy comes from sources known as electromotive forces (e.m.f.s) which transfer charges from a lower to a higher potential thus producing a potential *rise*. More about this in Programme 7.

30

ANALOGIES BETWEEN THE ELECTROSTATIC FIELD AND THE CONDUCTION FIELD

In Programme 3, Frame 13 we obtained an expression for the capacitance of a parallel plate capacitor in terms of its physical dimensions and the permittivity of the dielectric. In Frame 17 of this programme we obtained an expression for the conductance of the material between parallel plates in terms of its physical dimensions and the conductivity of the material. We obtained expressions for the capacitance (Programme 3, Frame 28) and leakage resistance (Frame 19 of this programme) of a coaxial cable in terms of its physical dimensions and the field constants.

A careful study of the derivation of these expressions will reveal the following analogous quantities in the two field systems.

Electrostatic field quantity		*Conduction field quantity*	
Electric charge	Q	Electric current	I
Electric flux density	D	Electric current density	J
Permittivity	ε	Conductivity	σ
Electric field strength	E	Electric field strength	E
Electric potential difference	V	Electric potential difference	V
Capacitance	C	Conductance	G

A useful consequence of these analogies is that for a given arrangement of conductors an expression for capacitance can be converted into an expression for conductance simply by replacing ε with σ (and vice versa).

31

Which of the following statements are true?

(a) The equivalent resistance of three 9 Ω resistors connected in series is 27 Ω.
(b) The equivalent resistance of three resistors connected in parallel is given by $1/R_1 + 1/R_2 + 1/R_3$.
(c) If a number of resistors are connected in parallel, the equivalent resistance is always less than the lowest individual resistance.
(d) The power dissipated in a resistance R when a current I flows through it is given by $I^2R/2$ (W).
(e) The unit of temperature coefficient of resistance is the °C.
(f) The algebraic sum of the currents meeting at a node is zero.
(g) Kirchhoff's second (voltage) law applies only to closed circuits.
(h) The vector J is the conduction field analogue of D.
(i) Charge has no analogous quantity in the conduction field.
(j) Resistivity and permittivity are field analogues.

32

> True: a, c, f, g, h

(b) The equivalent resistance is the *reciprocal* of this.   (Frame 26)
(d) The power dissipated is just I^2R.   (Frame 26)
(e) The unit is $(°C)^{-1}$.   (Frame 27)
(i) Electric current is the analogue.   (Frame 30)
(j) Conductivity is the analogue of permittivity.   (Frame 30)

33

SUMMARY OF FRAMES 24–32

New quantities and units

Quantity	Symbol	Unit	Abbreviation
Electrical power	P	watt	W
Temperature coefficient of resistance	α	per degree of temperature	$(°C)^{-1}$

EQUATIONS AND LAWS

For a number of resistors in series $R_T = R_1 + R_2 + \ldots + R_n \ (\Omega)$

For a number of resistors in parallel $\dfrac{1}{R_T} = \dfrac{1}{R_1} + \dfrac{1}{R_2} + \ldots + \dfrac{1}{R_n} \ (S)$

Concerning variation of resistance with temperature

$$R = R_S[1 + \alpha_S(T - T_S)] \ (\Omega)$$

also
$$\frac{R_1}{R_2} = \frac{1 + \alpha_0 T_1}{1 + \alpha_0 T_2}$$

Kirchhoff's current law: the algebraic sum of the currents at a node is zero.
Kirchhoff's voltage law: the algebraic sum of the voltages (e.m.f.s and potential drops) around any *closed* circuit is zero.

34

EXERCISES

(i) Calculate the five possible values of resistance obtainable when three resistors of 5 Ω, 10 Ω and 20 Ω are connected in series or parallel or series–parallel in all possible ways. Each time use all three resistors.

(ii) The various combinations of resistances in question (i) are connected in turn to a direct voltage supply of 100 V. Calculate the power dissipated in each case.

(iii) A length of copper wire has a resistance of 150 Ω at 20°C. Its temperature coefficient of resistance at 0°C is 0.004 28/°C. Calculate its resistance at 40°C.

(iv) A 100 V battery having an internal resistance of 4 Ω is connected in parallel with a 90 V battery having a resistance of 3 Ω. A resistor of 50 Ω is connected in parallel with the batteries. Use Kirchhoff's laws to determine the current flowing in each of the three parallel branches. (*Hint*: let the batteries supply currents I_1 and I_2 then use (a) the current law to put the current in the resistor in terms of I_1 and I_2 and (b) the voltage law to set up two simultaneous equations in I_1 and I_2.)

(v) A cuboid block of conducting material has sides of length a, b and c. Its conductivity is σ. Given an expression for the resistance between each pair of opposite faces.
If the block were of a dielectric material, having permittivity ε, use the method of field analogies to obtain an expression for the capacitance between two metal plates attached to the sides measuring $a \times b$.

35

SHORT EXERCISES ON PROGRAMME 4

The numbers after the questions indicate the frames where the answers may be found.

1. Explain what constitutes an electric current and give its unit. (2)
2. Write down the relationship between current and charge. (3)
3. Distinguish between convection current and conduction current. (3)
4. Give the symbol for and the unit of current density. (4)
5. Write down the continuity equation. (9)
6. What is a tube of flux? (10)
7. Explain what is meant by an ohmic material. (11)
8. State Ohm's Law in terms of current density. (11)
9. What is the unit of resistivity? (12)
10. State Ohm's Law in terms of electric current. (14)
11. What is the unit of conductivity? (11)
12. Give an expression for the resistance of a conductor in terms of its dimensions and the resistivity of the material. (16)
13. What is the unit of conductance? (14)
14. Explain the meaning of the 'Joule effect'. (26)
15. Give the unit of electrical power. (26)
16. How much power is dissipated in a resistor of 20 Ω when it carries 5 A? (26)
17. Define the temperature coefficient of resistance. (27)
18. What is the unit of the temperature coefficient of resistance? (27)
19. State Kirchhoff's current law. (28)
20. What is the conduction field analogue of permittivity? (30)

36

ANSWERS TO EXERCISES

Frame 23

(i) 10 mA; (ii) 7.64 kA m^{-2}; (iii) 5×10^7 S m^{-1}; (iv) 0.332 Ω; (v) 1.07×10^{11} Ω; (vi) 3×10^{11} Ω; (vii) $(b-a)/4\pi\sigma ab$ (Ω), 2.653×10^{14} Ω; (viii) 9.091×10^{-8} Ω; (ix) 5.7×10^9 S, 5.7×10^7 S, 1.425×10^7 S.

Frame 34

(i) 2.857 Ω, 35 Ω, 11.6 Ω, 14 Ω, 23.3 Ω.
(ii) 3500 W, 285.7 W, 862 W, 714.3 W, 429.2 W.
(iii) 161.83 Ω; (iv) 2.2 A, 0.386 A, 1.814 A; (v) $a/\sigma bc$, $b/\sigma ca$, $c/\sigma ab$, $\varepsilon ab/c$.

Programme 5

STATIONARY MAGNETIC FIELDS 1
(IN A VACUUM)

1

INTRODUCTION

In Programmes 2 and 3 we considered electrostatic fields which we found were produced by charges at rest and in Programme 4 we saw that conduction fields are produced by charges in motion. We now begin a study of magnetic fields 'at rest' or 'steady' magnetic fields, and the question arises as to how they are produced. Before Oersted's discovery in 1820 that current-carrying conductors produced magnetic fields in their vicinity the only known magnetism was that of natural magnets. Early experimenters found that the effects of these were greatest near their ends (which were called poles) and that the two ends produced opposite effects. Although it was realised that poles always occurred in pairs. Michell had the idea of considering them separately and showed in 1750 that the inverse square law applied to the forces between them.

After Oersted's discovery, work by Ampere led him to suggest that all magnetism is the result of electric current and that even the magnetism of magnets is the result of circulating currents within the structure of the material of the magnet. Modern theory supports this view and some authors will not admit to the existence of a magnetic pole. However, the concept is quite useful and we shall use it where it seems appropriate to do so.

In this programme you will learn about:

- magnetic dipoles, current loops and their equivalence
- magnetic pole strength and magnetic moment
- magnetic flux density
- magnetic lines of force (or flux lines or streamlines)
- magnetic field strength
- the inverse square law applied to magnetic poles
- magnetic potential and magnetic potential gradient
- Ampere's circuital law (the magnetic circuit law; the work law)
- toroidal and solenoidal coils
- the Biot–Savart law.

When you have completed this programme you should be able to:

- calculate the forces between magnetic poles, between magnets, between current loops and between combinations of these
- calculate the magnetic potential at a point or the magnetic potential difference between points in a magnetic field
- calculate the magnetic field strength and the magnetic flux density at a point due to magnetic poles, straight current-carrying conductors and current-carrying coils of various shapes.

2

MAGNETIC LINES OF FORCE

In order to picture the form of a magnetic field, we again use the concept of a line of force just as we did for electrostatic fields and conduction fields. A magnetic line of force (or flux line or streamline) is a continuous line whose direction at any point is the direction of the magnetic field at that point.

The form of the field due to a bar magnet is shown in the diagram (a) below. Oersted showed that the field due to a straight current-carrying conductor takes the form of concentric circles as given in diagrams (b) and (c).

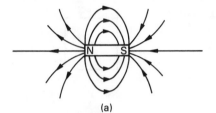

(a)

current out of paper

(b)

current into paper

(c)

3

MAGNETIC DIPOLES

As was mentioned in the introductory frame, the inverse square law applies to magnetic poles and the force between two poles of strength p_1 and p_2 separated by a distance d is given by

$$F = \frac{Kp_1p_2}{d^2} \qquad \text{where } K \text{ is a constant}$$

However, it is impossible to obtain 'a magnetic pole'. If a magnet is divided into two parts we are left with two magnets, not two poles. The smallest indivisible particle of magnetism is therefore a magnetic dipole.

A magnetic dipole may also be defined as a pair of equal and opposite poles separated by a distance which is small compared with the distance to the point at which we wish to measure its effects.

One of the earliest experimenters in magnetism, Gilbert (about 1600), found that a freely suspended magnet always comes to rest with one end pointing to the North pole of the earth. This end he called North and the other end is called South. He found that poles of the same sort repelled one another while poles of opposite polarity attracted one another.

4

CURRENT LOOPS

A current loop is defined as the current flowing in a 'small' circuit. 'Small' means small compared with the distance to the point at which we wish to measure its effects. The effective current in a circuit consisting of more than one turn is the current in one turn multiplied by the number of turns.

THE EQUIVALENCE BETWEEN CURRENT LOOPS AND MAGNETIC DIPOLES

The diagrams show a magnetic dipole (of strength p and separation $2l$) and a current loop (of area A carrying a current I). The line YY is the same in each diagram so that the magnetic dipole can be considered to be at the centre of the current loop and perpendicular to its plane. Visualise the loop as being formed of the two conductors in (b) and (c) of the diagram in Frame 2.

It is found that the magnetic behaviour of the dipole is identical with that of the loop if $\mu_0 IA$ for the loop is equal to $2lp$ for the dipole where μ_0 is a constant, called the permeability of free space. As we shall see in the next programme, its unit is the Henry per metre (H m^{-1}). In the SI it has the value $4\pi \times 10^{-7}$. For the dipole, $2lp = m$ (the *moment* of the dipole).

What further information must be given for complete equivalence?

5

The direction of the current in relation to the polarity of the dipole

The sense of the dipole is such that a right-hand screw would move towards the North end of the magnet when turned in the direction of the current.

This is illustrated in the diagrams above and this 'end rule' is easy to remember using the following diagrams in which the clockwise current is drawn in the shape of an S(outh) and the anticlockwise current is drawn in the shape of an N(orth).

circular coil   square coil

6

When a magnet is freely suspended, it comes to rest in a preferred direction with its north polar end pointing to the north pole of the earth (assuming that there are no other magnetic fields in the vicinity). If the magnet is then deflected from this position through an angle θ, it experiences a torque tending to reduce θ which is proportional to $\sin \theta$. For a given θ and a particular magnet, the magnitude of the torque varies with the location of the magnet because the strength of the earth's field varies from place to place. To gain a measure of the strength of the field at a particular place, we use a vector quantity called *Magnetic Flux Density*, symbol *B*, which is proportional to the torque.

7

THE UNIT OF MAGNETIC FLUX DENSITY

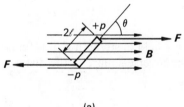   

(a)   (b)

The dipole shown in diagram (a) experiences a torque given by
    $T = $ (the force on the poles × the distance between the poles at right angles
       to the direction of the force)
    $= F \times 2l \sin \theta$ Newton metre (N m)
The force *F* will depend on the strength of the poles p and the strength of the field (*B*) in which they find themselves. Thus
$$T \propto p\,B\,2l \sin \theta$$
$$\propto m\,B \sin \theta \qquad \text{since } m = 2lp$$
   Remembering that, for equivalence between magnetic dipoles and current loops, $2lp\,(= m)$ for the dipole must equal $\mu_0 I A$ for the current loop, give an expression for the torque on the equivalent current loop shown in diagram (b).

8

$$T \propto IAB \sin \theta$$

Since μ_0 is a constant, it disappears in the proportionality sign. The SI unit of B is chosen so as to make the constant of proportionality equal to unity, so that $T = IAB \sin \theta$ (N m).

The SI unit of B so chosen is the tesla (T) which is equal to 1 weber per square metre. To gain some feel for magnitudes try to remember that 1 T is a very strong field and the earth's field is of the order of a few tens of μT.

What is the relationship between magnetic flux density B, and magnetic flux the symbol for which is Φ?

9

$$\Phi = BA, \text{ where } A \text{ is the area over which } B \text{ acts}$$

The unit of Φ is thus the weber (Wb) (weber metre^{-2} × metre^2).

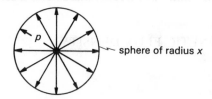
sphere of radius x

If we consider a pole of strength p to be producing a field of strength B, it is found experimentally that the magnitude of B varies as the inverse square of the distance from the pole.

At a distance x, $\qquad B_x = \dfrac{Kp}{x^2} \qquad$ (where K is a constant) $\hfill$ (i)

If we imagine a sphere of radius x to be drawn around p with p as its centre, then the flux crossing this sphere is given by the flux density at x (B_x) multiplied by the surface area of the sphere so that $\Phi = B_x \times 4\pi x^2$ (Wb). But $B_x x^2 = Kp$ from equation (i) above.

$\therefore \quad B_x \times 4\pi x^2 = 4\pi Kp$ and this is the flux produced by a pole of strength p.

The flux produced by a pole of strength 1 is therefore $4\pi K \times 1$ (Wb).

Unit pole is defined as that producing 1 Wb of flux.

What, then, is the value of K in equation (i)?

Since $4\pi K = 1$ Wb, then $K = 1/4\pi$

It follows that $B_x = p/4\pi x^2$ (compare this with $D_x = q/4\pi x^2$). Because a pole of unit strength produces one unit of flux, the same unit is used for both so that the unit of pole strength is the weber (Wb).

So what is the unit of magnetic moment?

11

The weber metre (Wb m)

MAGNETIC FIELD STRENGTH

We have seen that the torque on a dipole is given by $T = IAB \sin \theta$. Putting $m = \mu_0 IA$ we have that

$$T = \frac{m}{\mu_0} B \sin \theta \text{ (N m)}.$$

It is found to be convenient to introduce a vector called magnetic field strength, symbol H, whose unit we shall discover in Frame 19 to be the ampere per metre $(A \ m^{-1})$ and which is the magnetic field analogue of electric field strength (the E-vector). The vector H has the same direction as B and has a magnitude of B/μ_0 in free space so that $H = B/\mu_0$ and $B = \mu_0 H$.

Substituting into the above equation for torque, we write

$$T = mH \sin \theta$$

$$= 2lpH \sin \theta \tag{i}$$

The torque T is the force on the poles multiplied by the distance between the poles at right angles to the force, i.e.

$$T = F \times 2l \sin \theta \tag{ii}$$

Thus $\qquad\qquad F \times 2l \sin \theta = 2lpH \sin \theta \qquad\qquad$ from (i) and (ii)

Dividing both sides by $2l \sin \theta$, we have

$$F = pH \text{ (N)}$$

What is the analogous expression in the electrostatic field?

12

$$\boxed{F = QE}$$

THE FORCE BETWEEN TWO POLES OF STRENGTH p_1 AND p_2 SEPARATED BY A DISTANCE r

In Frame 10 we saw that the magnetic flux density at a distance x from a pole of strength p is given by

$$B_x = \frac{p}{4\pi x^2} \text{ (T)}$$

The flux density at a distance r from a pole of strength p_1 is therefore

$$B_r = \frac{p_1}{4\pi r^2} \text{ (T)}$$

The magnetic field strength at a distance r from the pole is

$$H_r = \frac{p_1}{4\pi r^2 \mu_0} \text{ (A m}^{-1})$$

If a pole of strength p_2 is placed in this field it will experience a force

$$F = p_2 H_r = \frac{p_2 p_1}{4\pi r^2 \mu_0} \text{ (N)}$$

This is the inverse square law.

Will the force be one of attraction or one of repulsion?

13

> Repulsion if p_1 and p_2 are both positive (North poles) or both negative (South poles); attraction if one is positive and the other is negative.

Example

Calculate the force between two magnets placed in the same straight line with a distance of 20 cm between their mid-points and with unlike poles adjacent. The magnets are each 5 cm long and have poles of strength 50×10^{-9} Wb. Remember, $\mu_0 = 4\pi \times 10^{-7}$ H m^{-1}.

The solution is given on the next page but try it yourself first. You will need to use the inverse square law and the Principle of Superposition.

Solution

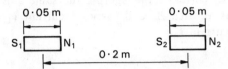

There will be forces of attraction between N_1 and S_2 and between N_2 and S_1. There will be forces of repulsion between N_1 and N_2 and between S_1 and S_2. We can use the inverse square law to calculate each of these four forces and then, using the Principle of Superposition, add them in order to obtain the resultant force between the magnets. In magnitude:

force between N_1 and S_2, $F_1 = \dfrac{+50 \times 10^{-9} \times (-50) \times 10^{-9}}{4\pi \times (0.15)^2 \times (4\pi \times 10^{-7})} = -7.04 \times 10^{-9} \, \text{N}$

force between N_1 and N_2, $F_2 = \dfrac{+50 \times 10^{-9} \times (+50) \times 10^{-9}}{4\pi \times (0.2)^2 \times (4\pi \times 10^{-7})} = +3.96 \times 10^{-9} \, \text{N}$

force between S_1 and S_2, $F_3 = \dfrac{(-50) \times 10^{-9} \times (-50) \times 10^{-9}}{4\pi \times (0.2)^2 \times (4\pi \times 10^{-7})} = +3.96 \times 10^{-9} \, \text{N}$

force between S_1 and N_2, $F_4 = \dfrac{(-50) \times 10^{-9} \times (+50) \times 10^{-9}}{4\pi \times (0.25)^2 \times (4\pi \times 10^{-7})} = -2.53 \times 10^{-9} \, \text{N}$

The resultant force between the two magnets is $F_1 + F_2 + F_3 + F_4$ and because all four forces act along the same straight line, the sum is a simple algebraic sum. If we let the resultant force be F_R then

$$F_R = (-7.04 + 3.96 + 3.96 - 2.53) \times 10^{-9} \, \text{N} = -1.65 \times 10^{-9} \, \text{N}$$

The negative sign indicates that the resultant force is attractive because it is equivalent to a force between poles of opposite polarity.

Could these two magnets be considered to be dipoles?

15

No, because the distance at which we measured their effects is of the same order as the distance between their poles.

16

THE FORCE BETWEEN CURRENT LOOPS

The diagram shows two current loops sharing the same axis and separated by a distance d. Loop 1 has an area A_1 and carries a current I_1 while loop 2 has an area A_2 and carries a current I_2.

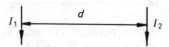

To determine the force between these current loops, we replace them by their equivalent dipoles and find the force between them using the inverse square law. The equivalent dipoles are shown below.

As for the example in Frame 13, there will be four forces acting and the resultant will be given by

$$F = \frac{p_1 p_2}{4\pi\mu_0}\left[-\frac{1}{(d-l_1-l_2)^2} + \frac{1}{(d-l_1+l_2)^2} + \frac{1}{(d+l_1-l_2)^2} - \frac{1}{(d+l_1+l_2)^2}\right] \text{(N)}$$

$$= \frac{p_1 p_2}{4\pi\mu_0}\left[-(d-l_1-l_2)^{-2} + (d-l_1+l_2)^{-2} + (d+l_1-l_2)^{-2} - (d+l_1+l_2)^{-2}\right] \text{(N)}$$

$$= \frac{p_1 p_2}{4\pi\mu_0 d^2}\left[-\left(1-\frac{l_1+l_2}{d}\right)^{-2} + \left(1-\frac{l_1-l_2}{d}\right)^{-2} + \left(1+\frac{l_1-l_2}{d}\right)^{-2} - \left(1+\frac{l_1+l_2}{d}\right)^{-2}\right] \text{(N)}$$

Each of the four terms inside the square brackets may be expanded in accordance with either $(1-x)^{-2} = 1 + 2x + 3x + 4x + \ldots$
or $\qquad (1+x)^{-2} = 1 - 2x + 3x - 4x + \ldots$

in which $\qquad\qquad x = \dfrac{l_1+l_2}{d} \quad$ or $x = \dfrac{l_1-l_2}{d}.$

Because $d \gg l_1 + l_2$ then we need only expand to the first three terms to get a sufficiently accurate result.

Try doing this now and thence obtain an expression for the resultant force between the equivalent dipoles.

17

$$F = \frac{p_1 p_2}{4\pi\mu_0 d^2}\left[6\left(\frac{l_1 - l_2}{d}\right)^2 - 6\left(\frac{l_1 - l_2}{d}\right)^2\right](\text{N})$$

Expanding $(l_1 - l_2)^2$ and $(l_1 + l_2)^2$, this becomes

$$F = \frac{p_1 p_2}{4\pi\mu_0 d^2}\left[\frac{24 l_1 l_2}{d^2}\right](\text{N}) \text{ which may be written as } \frac{6(2 l_1 p_1)(2 l_2 p_2)}{4\pi\mu_0 d^4}(\text{N})$$

By putting $2 l_1 p_1 = \mu_0 I_1 A_1$ and $2 l_2 p_2 = \mu_0 I_2 A_2$ and since $\mu_0 = 4\pi \times 10^{-7}$, we get

$$F = \frac{6 \times 10^{-7} I_1 I_2 A_1 A_2}{d^4}(\text{N}).$$

18

The *magnetic potential difference* between two points in a magnetic field is defined as the work required to move a unit positive pole between them. The symbol for magnetic potential is A and its unit is the ampere.

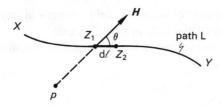

In the diagram, X and Y are two points lying in a magnetic field produced by the pole p. Z_1 and Z_2 are two points separated by a small distance, dl, on an arbitrarily chosen path (L) between X and Y.

The work which must be done by an external force in moving a unit positive pole from Z_1 to Z_2 is $F \cos \theta\, dl$ (J).

Since $F = pH$ and in this case $p = 1$, then the work required is $H \cos \theta\, dl$. By definition this is the potential by which point Z_1 exceeds that of point Z_2.

Using your experience of dealing with potential in the electrostatic field, give an expression for the potential difference between points X (A_x) and point Y (A_y) in terms of a suitable line integral.

19

$$A_x - A_y = -\int_L H \cos \theta \, dl \ (\text{A})$$

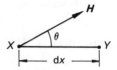

X and Y are two points separated by a small distance along the X-axis. The magnetic potential by which point Y exceeds that of point X is given by:

$dA = -H \cos \theta \, dx = H_x \, dx$ where H_x is the component of H along the X-axis.

$\therefore \quad H_x = -dA/dx$. The unit of H is thus the ampere per metre (A m^{-1}).

By analogy with the electrostatic field, what is dA/dx called?

20

Magnetic potential gradient in the X direction

In general, the component of magnetic field strength in any direction is the rate at which the magnetic potential falls in that direction.

Let us now obtain an expression for the magnetic potential at a point R, distant r from a pole of strength p.

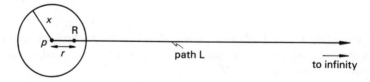

Consider a sphere of radius x to be drawn around the pole. The magnetic flux density at the surface of this sphere is given by

$$B_x = \frac{p}{4\pi x^2} \text{ and the magnetic field strength there is } H_x = \frac{p}{4\pi x^2 \mu_0}.$$

If the potential is taken to be zero at infinity, give an expression for the potential at the point R.

$$A_R = \frac{p}{4\pi\mu_0 r}(A)$$

This is obtained as follows:

$$A_R = -\int_L H_x\,dx = -\int_L \frac{p}{4\pi\mu_0 x^2}\,dx = -\frac{p}{4\pi\mu_0}\left[\frac{1}{-x}\right]_\infty^r = \frac{p}{4\pi\mu_0 r}(A)$$

We can use this result together with the Principle of Superposition to obtain an expression for the potential at a point due to a magnetic dipole.

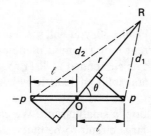

$$A_R = \frac{1}{4\pi\mu_0}\left[\frac{p}{d_1} - \frac{p}{d_2}\right](A)$$

Since, by the definition of a dipole, $r \gg 2l$

then $\qquad\qquad d_1 \approx r - l\cos\theta \qquad$ and $\qquad d_2 \approx r + l\cos\theta$

so that $\qquad\qquad A_R = \frac{p}{4\pi\mu_0}\left[\frac{1}{r - l\cos\theta} - \frac{1}{r + l\cos\theta}\right](A)$

$$= \frac{p}{4\pi\mu_0} \times \frac{2l\cos\theta}{r^2 - l^2\cos^2\theta}\,(A)$$

$$= \frac{p}{4\pi\mu_0} \times \frac{2l\cos\theta}{r^2} \qquad \text{since } r^2 \gg l^2\cos^2\theta$$

and finally $\qquad\qquad A_R = \frac{m\cos\theta}{4\pi\mu_0 r^2} \qquad \text{since } m = 2lp.$

What would be the magnetic potential at R due to the equivalent current loop?

22

$$A_R = IA\cos\theta/4\pi r^2$$

This is obtained simply by replacing m/μ_0 by IA where A is the loop area.

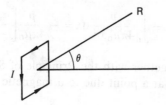

Referring to Programme 1, Frame 30, you will see that $A\cos\theta/r^2$ is the solid angle, Ω, subtended at R by the current loop (a small plane area). Thus $A_R = I\Omega/4\pi$.

Now 4π is the *total* solid angle subtended at R (Programme 1, Frame 27) so that $A_R = I \times$ (the fraction of the total solid angle at R which is subtended by the current loop).

This result is for a current loop, a small circuit. We can however extend the result to any circuit by imagining a large circuit to be made up of a large number of small circuits as shown in the diagram below.

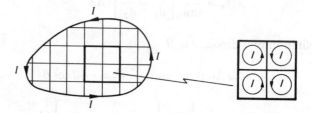

The arrangement is equivalent to a single large circuit because, except for the periphery (the large circuit) all the other currents cancel as they flow in opposite directions in adjacent circuits.

Each of the small circuits will subtend a solid angle at some point R at which we wish to determine the potential, and by the Principle of Superposition we have

$$A_R = \frac{I}{4\pi}\sum\Omega$$

Now move on to the next frame and decide which of the ten statements are true.

23

(a) The smallest indivisible particle of magnetism is the magnetic dipole.
(b) A current loop is a circular coil carrying a small current.
(c) The effective current in a current loop is NI where N is the number of turns in the loop and I is the current flowing in it.
(d) For equivalence between a current loop and a magnetic dipole, $\mu_0 IA$ must equal $2lp$ where all the symbols have their usual meaning.
(e) The nearer end of a coil carrying a clockwise current exhibits north (positive) polarity.
(f) Parallel lines of force in the same direction repel.
(g) The unit of magnetic flux density is the Weber.
(h) A magnetic flux density of 1 T represents a very strong magnetic field.
(i) A pole of strength p in a field of strength H experiences a force $F = pH$.
(j) The unit of magnetic potential is the ampere per metre.

24

True: a, c, d, f, h, i

(b) A current loop is a current flowing in a 'small' circuit.   (Frame 4)
(e) It exhibits south polarity.   (Frame 5)
(g) The unit is the tesla (weber per metre squared).   (Frame 8)
(j) The unit is the ampere.   (Frame 18)

25

SUMMARY OF FRAMES 1–24

Quantity	Symbol	Unit	Abbreviation
Magnetic pole strength	p	Weber	Wb
Magnetic moment	m	Weber-metre	Wb m
Magnetic flux density	B	Tesla	T
Magnetic field strength	H	Ampere per metre	$A\,m^{-1}$
Magnetic potential	A	Ampere	A
Magnetic potential gradient	H	Ampere per metre	$A\,m^{-1}$

Formulae:

$$B = \mu_0 H; \quad F = pH; \quad \mu_0 IA = 2lp.$$

26

EXERCISES

(i) Calculate the magnetic flux density and the magnetic field strength at a point 20 cm from a pole of 50 nWb.

(ii) Calculate the magnitude of the force on a pole of strength 30 nWb when it is placed in a magnetic field of density 15 mT.

(iii) Find the force of attraction between two magnets placed in the same straight line with a distance of 18 cm between their centres. One has a pole strength of 60 nWb and a length of 12 cm while the other has a pole strength of 45 nWb and a length of 6 cm.

(iv) Calculate the magnetic potential at a point $30 \angle 30°$ cm from the centre of a magnet having poles of strength 20 nWb separated by 1 cm.

27

AMPERE'S CIRCUITAL LAW

This very important law is used for calculating H and hence B at points in magnetic fields. It is also often referred to as 'the magnetic circuit law'.

In order to derive the law, we need to bear in mind the definition of magnetic potential difference (dA) between two points joined by a path L:

$$dA = - \int_L H \cos \theta \, dl$$

(Frame 19) and that the magnetic potential at a point P due to a current carrying circuit is given by $A_P = I\Omega/4\pi$ where Ω is the solid angle subtended at the point by the circuit (Frame 22).

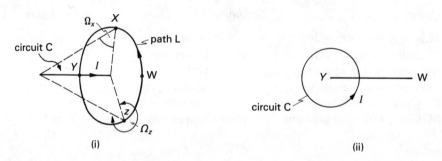

(i) (ii)

Diagram (i) shows a circuit, C, carrying a current I. A plan view of the arrangement is shown in diagram (ii). Let us now obtain an expression for the potential changes as we move a unit positive pole around a path L which *links* with the circuit, i.e. the path passes through the circuit.

If we start at point W, the solid angle subtended there by the circuit is zero because the point W is effectively outside a closed surface (see Programme 1, Frame 28). This can be taken as the datum for potential and is zero. The pole is now moved upwards against the direction of the field produced by the current so that the potential will be rising. At point X the solid angle subtended will be Ω_x where $0 < \Omega_x \times 2\pi$ and the potential will be $I\Omega_x/4\pi$.

The pole then moves downwards, still against the direction of the field so that the potential is still rising. At point Y the pole is in the plane of the circuit and the solid angle subtended is 2π steradian (Programme 1, Frame 28). When the pole reaches point Z the solid angle subtended by the circuit has increased to $\Omega_z (> 2\pi$ steradian) and the pole has in effect entered an open surface which now begins to close behind it. Finally the pole arrives back at point W and is inside a closed surface. The solid angle subtended at W by the circuit (now effectively a closed surface) is 4π steradian.

What magnetic potential difference has the pole moved through in one complete circuit of the path L?

28

$$\boxed{I\,(\text{A})}$$

Initially, $A = I \times 0/4\pi = 0$; finally, $A = I \times 4\pi/4\pi = I$. By definition, this is

$$-\oint_L H \cos \theta \, \mathrm{d}l \qquad \text{so that} \quad -\oint_L H \cos \theta \, \mathrm{d}l = I.$$

The negative sign indicates that the pole is moving against the direction of the field so that the potential is rising.

If the movement had taken place in the opposite direction, then we would have

$$\oint_L H \cos \theta \, \mathrm{d}l = I.$$

Note that if path L is traversed again, the potential at point W will change by a further I. In fact, every time path L is traversed, the potential at any point on it changes by I. Remember that the change in *electric* potential around a closed path is always zero.

What would be the result if the circuit consisted of N turns rather than one?

29

$$\oint_L H \cos \theta \, dl = NI$$

In general, we write

$$\oint_L H \cos \theta \, dl = \sum I.$$

This is Ampere's circuital law and is read 'the line integral of H around any closed path is equal to the total current linked with that path'. Note that the path must link the circuit otherwise the motion would be against the direction of the field for half the circuit and in the direction of the field for the other half, so that the net result would be zero.

30

Example

Calculate the magnetic flux density at a point perpendicularly 18 cm from a very long straight conductor carrying a current of 10 A.

Solution

To find the magnetic flux density we use $B = \mu_0 H$ which means we need an expression for H. Ampere's circuital law relates H and I.

The magnetic field due to a long straight conductor takes the form of concentric circles whose centres are on the axis of the conductor.

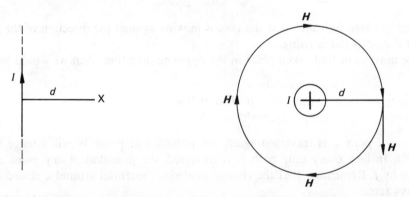

Can you suggest a suitable path which we could use to apply Ampere's law?

31

A circular line of force of radius 18 cm

Applying Ampere's law to the path we have

$$\oint_L H \cos \theta \, dl = \sum I.$$

The H-vector is everywhere tangential to the path and so $\cos \theta = 1$. The magnitude of H at all points on this path is the same by symmetry. The current linked with the path is I (10 A).

$$\therefore \quad H \oint_L dl = I \quad \text{and} \quad H \times 2\pi d = I$$

$$H = I/2\pi d \text{ (A m}^{-1}) \quad \text{and} \quad B = \mu_0 I/2\pi d \text{ (T)}$$

Putting in the values $I = 10$ A, $d = 0.18$ m and $\mu_0 = 4\pi \times 10^{-7}$, $B = 11.1 \ \mu\text{T}$.

The next frame gives another example of the use of Ampere's law.

32

THE MAGNETIC FIELD OF A TOROIDAL COIL

A toroidal coil is a coil wound on a circular former which may have a cross-sectional area of any shape. If the coil is closely wound, the magnetic field will be represented by concentric circles.

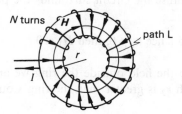

To investigate the magnetic field within the volume of the ring we apply Ampere's law to any path such as L:

$$\oint_L H \cos \theta \, dl = \sum I$$

Because H is constant in magnitude around the path by symmetry and is acting in the direction of the path at every point on it, then $\cos \theta = 1$ and H may be taken through the integral sign. Also the path links all of the N turns of the coil. Bearing this in mind, give an expression for H.

$$\boxed{H = NI/2\pi r \ (\text{A m}^{-1})}$$

This comes from applying Ampere's law to path L which gives

$$H\oint_L dl = NI \qquad \text{so that} \qquad H \times 2\pi r = NI \qquad \text{and} \qquad H = NI/2\pi r.$$

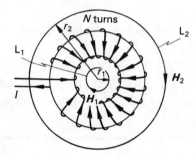

To investigate the magnetic field in the space in the 'hole' of the ring we apply Ampere's law to any path such as L_1 for which r_1 is greater than zero but less than the inner radius of the ring.

$$H_1 \int_{L_1} dl = 0 \text{ because the circuit surrounds the path but does not link it}$$

$$\therefore \quad H_1 \times 2\pi r_1 = 0 \text{ and since } r_1 \neq 0, \text{ then } H_1 = 0$$

Similarly to investigate the field outside the ring we apply Ampere's law to any path such as L_2 for which r_2 is greater than the ring's outer radius.

$$H_2 \int_{L_2} dl = 0 \text{ because the path surrounds the current but does not link it}$$

$$\therefore \quad H_2 \times 2\pi r_2 = 0 \text{ and since } r_2 \neq 0, \text{ then } H_2 = 0$$

The magnetic field of a toroidal coil is therefore confined within the volume of the ring.

What is the magnitude of the magnetic field strength along the mid-line of a toroidal coil of 550 turns wound on a ring having an inside diameter of 15 cm and an outside diameter of 22 cm? The current in the coil is 5 A.

$$\boxed{4732 \text{ A m}^{-1}}$$

This is obtained from the formula in the box at the top of Frame 33 with $N = 550$, $I = 5$ A and $r = 9.25$ cm.

THE MAGNETIC FIELD OF A VERY LONG SOLENOID

A solenoid is a coil wound on a straight former. It could be considered to be a toroidal coil having infinite radius in which case we can make use of the results obtained for such a coil.

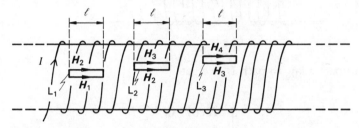

Along the mid-line of the solenoid, the magnetic field strength is given by

$$H_1 = NI/2\pi r \text{ (Frame 33)}$$

This is the result obtained for a toroidal coil with $r = \infty$ but we cannot put $r = \infty$ because this would give a meaningless result mathematically, so we put it in the form $H_1 = I \times$ the number of turns per unit length of the coil.

To examine the field along a line close to the mid-line, we apply Ampere's law to a rectangular path such as L_1:

$$\int_{L_1} H \cos \theta \, \mathrm{d}l = \sum I$$

Now along the longer sides $\theta = 0$ so that the corresponding $\cos \theta = 1$. Along the shorter sides $\theta = 90°$ so that $\cos \theta = 0$. Also, the current linking the path is zero (the current surrounds the path without linking it). Putting all this information into the formula we get

$$H_1 l - H_2 l = 0$$

and so
$$H_1 = H_2$$

By induction, i.e. by applying Ampere's law to successive paths such as L_2, L_3 and so on, we find that $H_3 = H_2$, $H_4 = H_3$ and so on. In other words, the magnetic field strength everywhere within the volume of the solenoid is the same and lines of force within it are parallel, straight lines.

How can we investigate the field outside the solenoid?

35

By applying Ampere's law to a path, part of which encroaches on the space outside the solenoid.

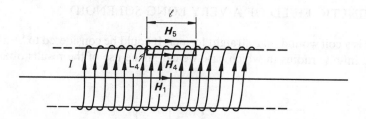

Applying Ampere's law to path L_4 we get

$$\oint_{L_4} H \cos \theta \, \mathrm{d}l = \sum I$$

Here, $H_4 = H_1$ (from considerations in the previous frame) and the current linked with the path is $I \times$ (the number of turns per unit length) $\times$ the length of the path, i.e. $I \times N/2\pi r \times l$.

$$\therefore \quad H_1 l - H_5 l = I \times N/2\pi r \times l$$

$$H_1 - H_5 = I \times N/2\pi r$$

But $H_1 = I \times N/2\pi r$ (see Frame 34) so that $H_5 = 0$.

There is therefore no magnetic field outside the volume of an infinitely long solenoid.

You will have noticed that in obtaining expressions for the magnetic field strength due to a straight wire carrying a current and due to a solenoid, we have stipulated that they must be infinitely long. Ampere's law gives correct answers in such cases. Shorter wires or solenoids have end effects which the method overlooks. However, reasonably accurate answers are obtained using this method in those cases where the length of the straight wire or of the solenoid is large compared with the distance to the point where we wish to determine the magnetic field strength or magnetic flux density.

In the next frame we meet another method for determining magnetic fields which does not have the same restriction as Ampere's law.

36

THE BIOT–SAVART LAW

In the same year that Oersted showed that electric currents produced magnetic fields in their vicinity, Biot and Savart described the effect quantitatively and developed what has become known as the Biot–Savart law.

It is used for calculating magnetic field strengths and flux densities due to various systems of current-carrying circuits and may be stated as follows: 'the magnetic field strength at a point due to a current flowing in a circuit is the same as if each current element in the circuit contributed a vector

$$dH = \frac{I \, dl \sin \theta}{4\pi x^2}$$

where θ is the angle between the current element and the line joining the current element to the point and x is the distance from the current element to the point.

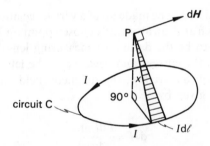

$I \, dl$ is one current element which is defined as the current, I, flowing in a very short length, dl, of the circuit. The circuit, C, is made up of a very large number of these current elements and each one produces a dH at P. The total field strength at P due to the whole circuit is the vector sum of all of the components like dH which are contributed by all the current elements.

Is there anything else we need to know for complete specification of the resultant magnetic field strength at P?

37

| Yes, we need to know the direction of each dH |

This is at right angles to the plane containing the current element and the line joining the current element to the point (this plane is shown shaded in the diagram above) and clockwise about the direction of the current.

38

We shall now use the Biot–Savart law to obtain an expression for the magnetic field strength at a point P, perpendicularly distant r from a short straight wire. Of course, current cannot flow in a short wire which is isolated but it is important because square coils can be thought of as being made up of four straight wires. Hexagonal coils are made up of six, and so on.

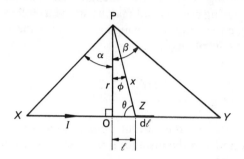

The wire XY is considered to be made up of a very large number of current elements $I\,dl$. One of these is shown at Z, an arbitrarily chosen point on XY. PO is perpendicular to XY and O is taken to be the datum for measuring length, positive being to the right in the direction of the current and negative to the left.

According to the Biot–Savart law, the magnetic field strength at P due to the current element at Z is given by

$$dH = \frac{I\,dl\sin\theta}{4\pi x^2}.$$

What is the direction of $d\mathbf{H}$?

39

At right angles to the plane of the paper and out of it

As Z takes all possible positions from point X at the extreme left-hand side of the wire to point Y at the extreme right-hand side, so l, θ and x will all vary. Since we are inevitably going to have to integrate something at some stage, it is important to put all these variables in terms of one of them or in terms of some other, more convenient variable. It so happens that in this case it is more convenient to put everything in terms of the angle ϕ, which varies from $-\alpha$ (when Z is at X) to $+\beta$ (when Z is at Y). It is $-\alpha$ because X is on the negative side of the datum O.

The next step, then, is to express everything in terms of ϕ.

Now $\quad l = r \tan \phi \quad$ so that $\quad \dfrac{\mathrm{d}l}{\mathrm{d}\phi} = r \sec^2 \phi \quad$ and $\mathrm{d}l = r \sec^2 \phi \, \mathrm{d}\phi$

also $\quad \sin \theta = \cos \phi$

and $\quad x/r = \sec \phi$

so that $\quad x = r \sec \phi.$

Putting these into the formula we get

$$\mathrm{d}H = \frac{I r \sec^2 \phi \, \mathrm{d}\phi \cos \phi}{4\pi r^2 \sec^2 \phi} = \frac{I \cos \phi \, \mathrm{d}\phi}{4\pi r} \ (\mathrm{A \ m}^{-1})$$

directed out of the plane of the paper and at right angles to it. This is the contribution to the total field strength at P from the current element at Z.

Now complete the problem to obtain the expression for the total magnetic field strength at P.

40

$$H_\mathrm{P} = \frac{I}{4\pi r} (\sin \beta + \sin \alpha) \ (\mathrm{A \ m}^{-1})$$
directed at right angles to and out of the paper

Here is the working:

The total field strength at P, $H_\mathrm{P} = \displaystyle\int_{-\alpha}^{+\beta} \frac{I}{4\pi r} \cos \phi \, \mathrm{d}\phi \ (\mathrm{A \ m}^{-1})$

$$= \frac{I}{4\pi r} \int_{-\alpha}^{+\beta} \cos \phi \, \mathrm{d}\phi = \frac{I}{4\pi r} [\sin \phi]_{-\alpha}^{+\beta}$$

$$= \frac{I}{4\pi r} (\sin \beta - \sin(-\alpha)) = \frac{I}{4\pi r} (\sin \beta + \sin \alpha)$$

What would be the result for an infinitely long wire?

41

$$H = I/2\pi r \ (\text{A m}^{-1})$$

For an infinitely long wire $\alpha = \beta = 90°$ so that $(\sin \beta + \sin \alpha) = 2$. Note that this is the same result as we obtained using Ampere's law.

Example

Calculate the magnetic field strength at a point perpendicularly 10 cm from the centre of a straight wire of length 30 cm carrying a current of 50 A.

Solution

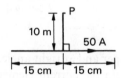

In this case, $\alpha = \beta = \tan^{-1}(15/10) = 56.31°$.

$$H_P = \frac{50}{4\pi 0.1}(2 \sin 56.31°) = \underline{66.21 \text{ A/m}}$$

H_P is directed at right angles to the plane of the paper and out of it.

42

THE MAGNETIC FIELD STRENGTH ON THE AXIS OF A SINGLE TURN CIRCULAR COIL

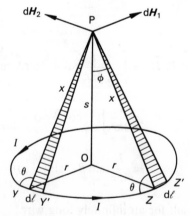

The coil has a radius r and carries a current I.

According to the Biot–Savart law the magnetic field strength at P, a point on the axis of the coil, due to the current element $I\,dl$ at Z is given by

$$dH_1 = \frac{I\,dl\sin\theta}{4\pi x^2} \text{ directed as shown}$$

As Z takes all possible positions around the circumference of the coil, x remains constant in magnitude and θ remains constant at $90°$ so that

$$dH_1 = \frac{I\,dl}{4\pi x^2} \text{ at right angles to the plane PZZ}'$$

Similarly, the magnetic field strength at P due to the current element at Y is given by

$$dH_2 = \frac{I\,dl}{4\pi x^2} \text{ at right angles to the plane PYY}'$$

What will be the direction of the resultant field strength at P?

43

Along the axis and vertically upwards

This is because every dH will have the same magnitude by symmetry. Each will have a component vertically upwards along the axis and a component at right angles to the axis. The components at right angles to the axis will cancel in pairs while the components along the axis will add.

The component along the axis is $dH\cos(90° - \phi)$

$$= \frac{I\,dl}{4\pi x^2}\cos(90° - \phi) = \frac{I\,dl}{4\pi x^2}\sin\phi \ (A\ m^{-1})$$

The resultant field strength at P is the sum of all these from all the current elements which go to make up the current in the circular coil. Since ϕ, I and x are all constants, this reduces to

$$H_P = \frac{I\sin\phi}{4\pi x^2}\sum dl$$

The sum of all the dl terms which go to make up the coil is the circumference of the coil, $2\pi r$, so that

$$H_P = \frac{I\sin\phi}{4\pi x^2}2\pi r \qquad \text{along the axis}$$

Obtain expressions for H_P in terms of (a) r and ϕ, (b) r and s.

44

$$\text{(a) } H_P = \frac{I \sin^3 \phi}{2r}; \qquad \text{(b) } H_P = \frac{Ir^2}{2(r^2 + s^2)^{3/2}} \text{ (A m}^{-1})$$

(a) is obtained by putting $x = r/\sin \phi$; for (b), put $x = (r^2 + s^2)^{1/2}$.

What is the magnetic field strength at the centre of a concentrated coil of N turns? (A 'concentrated' coil is one in which all the turns are considered to occupy the same space.)

45

$$H = NI/2r \text{ (A m}^{-1})$$

In this case the effective current is NI ampere and $\phi = 90°$, so that $\sin^3 \phi = 1$.

Example

Two single-turn circular coils share a common horizontal axis and are separated by a horizontal distance of 2.5 m. The radius of one of the coils is 50 cm and it carries a current of 5 A. The radius of the other coil is 25 cm and it carries a current I in such a direction that the magnetic field which it produces is in opposition to that produced by the current in the first coil.

Determine the value of I in order that the resultant magnetic field strength at the mid-point on the axis between the coils shall be zero.

Solution

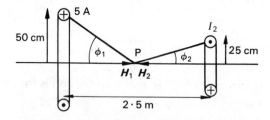

For the resultant field strength at P to be zero, $H_1 = H_2$ there.

The formula for the magnetic field strength on the axis of a circular coil was derived in Frame 43. Using this formula in the form (a) in the box in Frame 44 we have, for the required condition,

$$\frac{I_1 \sin^3 \phi_1}{2r_1} = \frac{I_2 \sin^3 \phi_2}{2r_2}$$

$\phi_1 = \tan^{-1}(0.5/1.25) = 21.8°$ and $\sin^3 \phi_1 = 0.051$
$\phi_2 = \tan^{-1}(0.25/1.25) = 11.3°$ and $\sin^3 \phi_2 = 0.0075$
$r_1 = 0.5$ m; $r_2 = 0.25$ m; $I_1 = 5$ A.
Solving the equation for I_2 gives $\underline{I_2 = 17 \text{ A}}$.

THE MAGNETIC FIELD OF A SHORT SOLENOID

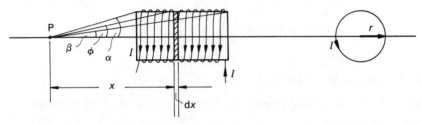

If we consider the solenoid to be uniformly, closely wound with circular coils, each one will be virtually at right angles to the axis and we can make use of the result obtained in Frame 43.

Let the coil have n turns per metre length so that a small length dx will have $(In\,dx)$ ampere flowing in it. This small section of the solenoid will behave like a circular coil for which the magnetic field strength at P is given by

$$dH = \frac{(In\,dx)\sin^3\phi}{2r} \text{ (A m}^{-1})$$

directed along the axis.

The solenoid is made up of a very large number of such sections. For the extreme left-hand section, $\phi = \alpha$ while for the extreme right-hand section $\phi = \beta$. For each section x will vary, too, so it will be as well to put x in terms of ϕ for the purpose of integration.

Now $x = r \cot \phi$ so that

$$\frac{dx}{d\phi} = -r \operatorname{cosec}^2 \phi$$

and $dx = -r \operatorname{cosec}^2 \phi \, d\phi$.

Substituting this into the formula we get

$$dH = \frac{In(-r \operatorname{cosec}^2 \phi \, d\phi) \sin^3 \phi}{2r}.$$

Now obtain an expression for the field strength at P due to the entire solenoid.

47

$$\boxed{H_P = \frac{nI}{2}(\cos \beta - \cos \alpha) \ (\text{A m}^{-1})}$$

acting along the axis

Here are the final steps in the working:

$$\mathrm{d}H = \frac{nI(-r \operatorname{cosec}^2 \phi \, \mathrm{d}\phi) \sin^3 \phi}{2r} = -\frac{nI \sin^3 \phi \, \mathrm{d}\phi}{2 \sin^2 \phi} = -\frac{nI \sin \phi \, \mathrm{d}\phi}{2}$$

$$H_P = -\frac{nI}{2} \int_\alpha^\beta \sin \phi \, \mathrm{d}\phi = -\frac{nI}{2} [-\cos \phi]_\alpha^\beta = \frac{nI}{2}[\cos \phi]_\alpha^\beta$$

and finally, putting in the limits, $H_P = \dfrac{nI}{2}(\cos \beta - \cos \alpha) \ (\text{A m}^{-1})$.

What is the magnetic field strength at the centre of a very long solenoid?

48

$$\boxed{H = nI \ (\text{A m}^{-1}) \text{ directed along the axis}}$$

This is because β becomes $0°$ and α becomes $180°$.

$$\therefore \qquad H = \frac{nI}{2}(\cos 0° - \cos(-180°)) = \frac{nI}{2}(2)$$

Note that this is the same result as we obtained using Ampere's law.

Example

Determine the magnetic flux density at the centre of a solenoid of length 20 cm and diameter 50 cm if it is uniformly wound with 1200 turns of wire carrying 3 A.

The solution is given in the next frame but try it yourself first.

49

Solution

Referring to the diagram in Frame 46, the point P in this case will be at the centre of the solenoid.

Thus $\beta = \tan^{-1}(25/10) = 68°$ and $\alpha = (180° - \beta) = 112°$.

We have that

$$B_P = \mu_0 H_P = \frac{\mu_0 nI}{2}(\cos \beta - \cos \alpha) \text{ along the axis}$$

with $I = 3$ A, $\mu_0 = 4\pi \times 10^{-7}$ and $n = 1200/0.2 = 6000$.

Substituting these values into the formula gives $B_P = 8.4$ mT along the axis.

Now decide which of the following statements are true.

(a) Ampere's circuital law can only be applied to infinitely long conductors.
(b) Ampere's circuital law may be expressed as $\oint_L H \cos \theta \, dl = \Sigma I$.
(c) A toroidal coil is a coil wound on a ring of circular cross-section.
(d) The magnetic field of a toroidal coil is confined within its volume.
(e) The field inside a toroidal coil consists of parallel straight lines.
(f) The field inside a long solenoid is everywhere of the same strength.
(g) Magnetic potential is exactly analogous to electric potential.
(h) The Biot–Savart law gives only the magnitude of the field strength whereas Ampere's law gives both the magnitude and direction.
(i) A current element is a current flowing in a very short piece of wire.
(j) The Biot–Savart law and Ampere's will only give identical results for the field strength due to a straight conductor if it is infinitely long.

50

| True: b, d, f, i, j |

(a) It can be applied to many other problems.   (Frames 30–35)
(c) The ring can be of any cross-section.   (Frame 32)
(e) The field is circular.   (Frame 35)
(g) Magnetic potential doesn't have a unique value.   (Frame 28)
(h) The magnitude and direction of H is given by the law.   (Frames 36 & 37)

51

SUMMARY OF FRAMES 27–50

Ampere's circuital law: the line integral around any closed circuit is equal to the current linked with that circuit;

$$\oint_L H \cos \theta \, dl = \sum I.$$

The Biot–Savart law: the magnetic field strength at a point due to a circuit carrying a current may be calculated by adding the contributions from all the current elements which go to make up the circuit. Each current element contributes

$$dH = \frac{I \, dl \sin \theta}{4\pi x^2}$$

at right angles to the plane containing $I \, dl$ and x, and clockwise about the current. Due to a straight wire:

$$H_P = \frac{I}{4\pi r}(\sin \beta + \sin \alpha)$$

Due to a solenoid:

$$H_P = \frac{nI}{2}(\cos \beta - \cos \alpha)$$

Note: n is the number of turns *per unit length*.
For a toroidal coil:

$$H = NI/2\pi r$$

52

EXERCISES

(i)   A ring having a square cross-section is uniformly wound with 500 turns of wire carrying a current of 2 A. The inner radius of the ring is 20 cm and its outer radius is 30 cm. Calculate the maximum and minimum values of flux density in the ring.

(ii) Calculate the magnetic field strength and flux density at a point perpendicularly 10 mm from a long straight conductor carrying 10 A.

(iii) Calculate the magnetic field strength at a point perpendicularly 10 cm from the mid-point of a straight wire 16 cm long carrying 100 A.

(iv) A solenoid of radius 10 cm and length 20 cm is uniformly wound with 200 turns of wire and carries a current of 5 A. Calculate the magnetic flux density on the axis and (a) at the ends of the coil, (b) in the middle of the coil.

53

SHORT EXERCISES ON PROGRAMME 5

The numbers in brackets refer to the frames where the answers may be found.

1. Define a magnetic dipole.   (3)
2. Define a current loop.   (4)
3. State the equivalence between current loops and dipoles.   (4)
4. What is the unit of magnetic pole strength?   (10)
5. Explain what is meant by 'magnetic moment'.   (4)
6. Give the unit of magnetic moment.   (11)
7. Give the unit of magnetic flux density.   (8)
8. Give the symbol for and the unit of magnetic flux.   (8, 9)
9. What is the relationship between the vector B and the vector H?   (11)
10. What is the unit of magnetic field strength?   (19)
11. Give the relationship between H, p and F.   (11)
12. State the inverse square law as applied to magnetic poles.   (12)
13. Define magnetic potential difference.   (18)
14. Give the unit of magnetic potential.   (18)
15. State mathematically the relationship between magnetic field strength and magnetic potential gradient.   (19)
16. State Ampere's circuital law.   (29)
17. What is the form of the field in a long solenoid?   (34, 35)
18. State the Biot–Savart law.   (36, 37)
19. Explain what is meant by a toroidal coil.   (32)
20. What is the form of the field due to a toroidal coil?   (32, 33)

54

ANSWERS TO EXERCISES

Frame 26

(i) 9.95×10^{-8} T, 7.92×10^{-2} A m^{-1};
(ii) 4.5×10^{-10} N;
(iii) 1.2×10^{-8} N;
(iv) 2.43×10^{-4} A.

Frame 52

(i) 1×10^{-3} T, 6.6×10^{-4} T;
(ii) 159.2 A m^{-1}, 2×10^{-4} T;
(iii) 99.4 A m^{-1};
(iv) 2.8×10^{-3} T, 4.45×10^{-3} T.

Programme 6

STATIONARY MAGNETIC FIELDS 2
(IN FERROMAGNETIC MEDIA)

1

In the previous programme we studied magnetic fields in 'free space' which meant in practice considering magnetic flux lines to be set up in air. We now consider the effects of the presence of material media and we find that in some ways, magnetic materials are similar to dielectric materials in the electric field. For example, polarisation takes place in magnetic materials under the influence of the H-vector just as it takes place in a dielectric material under the influence of the E-vector.

However there is one very important difference in that whereas dielectric materials are nearly always linear (i.e. D is proportional to E), a large group of magnetic materials is highly non-linear. Furthermore these materials behave in a manner which is dependent on their history.

In this programme, then, we will learn about:

- paramagnetic, diamagnetic and ferromagnetic materials
- magnetic domains and the magnetisation of iron
- saturation curves and hysteresis loops
- relative permeability
- incremental permeability
- 'hard' and 'soft' magnetic materials
- hysteresis loss
- magnetic circuits and their analogy with electric circuits
- reluctance and permeance
- magnetic leakage and fringing effects
- permanent magnets
- boundary conditions between different media in magnetic fields
- the law of magnetic flux refraction.

When you have studied this programme, you should be able to:

- distinguish between paramagnetic, diamagnetic and ferromagnetic materials
- describe the process of magnetisation of iron
- calculate the hysteresis loss associated with alternating magnetisation
- solve series and parallel magnetic circuit problems
- describe the characteristics of permanent magnet materials
- perform minimum volume calculations for permanent magnets
- apply the law of magnetic flux refraction.

2

CLASSIFICATION OF MAGNETIC MATERIALS

Virtually all materials are affected by magnetic fields, and according to their magnetic behaviour they are classified as 'paramagnetic' or 'diamagnetic'. In 1845 Faraday showed that, in a uniform magnetic field, rectangular samples of all materials align themselves with their longer axis parallel to the field. However, in a non-uniform magnetic field some materials (which he called *paramagnetic*) align themselves with their longer axis parallel to the field, but other materials (which he called *diamagnetic*) align themselves with their longer axis at right angles to the direction of the field.

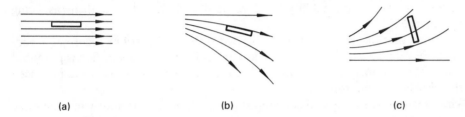

(a)   (b)   (c)

The diagram shows samples of materials situated in magnetic fields. In each case, decide whether the material is paramagnetic or diamagnetic.

3

> (a) Either – because the field is uniform.
> (b) Paramagnetic – because it is parallel with a non-uniform field.
> (c) Diamagnetic – because it is at right angles to a non-uniform field.

FERROMAGNETISM

All diamagnetic materials and most paramagnetic materials exhibit only very weak magnetic effects and most of them show such weak effects that they are referred to as being 'non-magnetic'. However only a vacuum is truly non-magnetic.

Within the group of paramagnetic materials is a sub-group having very strong paramagnetic properties. A typical example of a material in this sub-group is iron and for this reason such materials are called *ferro-magnetic*, 'ferro-' meaning 'of or containing iron'.

Which of the following materials do you think are ferromagnetic:

aluminium, copper, mild steel, nickel, silver?

4

> mild steel, nickel
> (aluminium is paramagnetic; the other two are diamagnetic)

Paramagnetic materials are more permeable than a vacuum (i.e. it is easier to establish a magnetic field in them) by a factor of up to 1.000 02 times. The ferromagnetic materials are more permeable by a factor of hundreds, thousands or more times! Diamagnetic materials are less permeable than a vacuum by a factor of 0.999 99 times. So far as we engineers are concerned, all but the ferromagnetics may be regarded as having the same permeability as a vacuum and we refer to them as being 'non-magnetic'.

It is found that the magnetic field associated with a coil carrying a current is very much increased if an iron bar is inserted into the coil. Since magnetic flux is produced by a number of ampere-turns, it seems that extra ampere-turns are being produced by the presence of the iron bar.

When a current-carrying coil is wrapped around a piece of iron or when the piece of iron is stroked with a permanent magnet, it becomes 'magnetised' and its magnetism is regarded as being due to tiny electric currents circulating within its molecules. The 'ampere-turns' of these currents add to those of the coil to produce a stronger field.

What do you think constitute the tiny circulating currents?

5

> The orbiting electrons in the atoms of the iron

THE MAGNETISATION OF IRON

The atoms of iron are arranged in groups called *domains*. Each domain is about one-tenth of a millimetre wide and contains of the other order of 10^{15} atoms whose magnetic axes are more or less parallel so that each domain is virtually a tiny magnet. In unmagnetised iron the magnetic axes of the domain are pointing randomly in various directions and there is no net effect in any particular direction. When a sample of iron is 'magnetised', the domains are rearranged so that their axes all point in the same direction, that of the magnetising field. The magnetising field then appears to be stronger because it is added to by the rearranged domains. When the magnetising field is removed there is some stability, some of the domains remaining in their new positions. In other words, the iron retains some of its magnetisation. Some materials (for example, soft iron) retain very little magnetisation once the magnetising field has been removed. Others (for example, hardened steel) retain a great deal and are used for permanent magnets (see Frame 51).

6

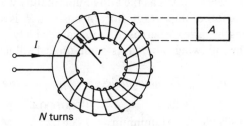

N turns

RELATIVE PERMEABILITY

If the ring shown in the diagram is of a non-magnetic material, the magnetic field strength H at radius r is given by

$$H = \frac{NI}{2\pi r} \quad \text{(A m}^{-1}) \text{ (see Programme 5, Frame 33)}$$

and the flux density there is given by

$$B = \mu_0 H \quad \text{(T)}$$

If the ring is of iron the flux will be much greater as the iron becomes magnetised. Let it be B_i.

Then $$\frac{\text{flux with an iron ring}}{\text{flux with a non-magnetic ring}} = \frac{B_i A}{BA} = \frac{B_i}{B}.$$

The ratio is called the relative permeability of the iron, symbol μ_r, and being a simple ratio, it has no units.

Since $$B = \mu_0 H$$

then $$B_i = \mu_r \mu_0 H$$

$$= \mu H$$

where $\mu = \mu_r \mu_0$ and is called the absolute permeability of the iron.
What are the units of μ?

7

T m A^{-1} (unit of B/unit of H). As we shall see in the next programme, this is equivalent to the Henry per metre.

μ_r is the magnetic field analogue of ε_r (relative permittivity) in the electrostatic field but it has a much wider range of values. For example, bismuth, a strong diamagnetic material, has $\mu_r = 0.99983$ whereas the alloy Supermalloy which contains 79% nickel has $\mu_r = 100000$. Another important difference is that ε_r is substantially constant as electric field strength varies whereas μ_r can vary widely with the strength of the magnetic field. The following table gives some typical values.

Diamagnetic	μ_r	Paramagnetic	μ_r	Ferromagnetic	μ_r
Bismuth	0.99983	Air	1.0000004	Nickel	600
Silver	0.99998	Aluminium	1.00002	Mild steel	2×10^3
Lead	0.999983	Palladium	1.0008	Silicon iron	7×10^3
Copper	0.999991			Purified iron	2×10^5

8

MAGNETISATION CURVE

The extra flux density as a result of the iron (Frame 4) is given by $B - \mu_0 H$ and is found to have a definite maximum for a given sample. No matter how much H is increased after a certain value, $B - \mu_0 H$ remains constant. This happens when all of the domains have been aligned in the direction of the magnetising field, and a state of magnetic saturation is said to have been reached. If the magnetic flux density B is plotted as a function of the magnetic field strength H, the so-called magnetisation curve or saturation curve is obtained. A typical curve is shown below.

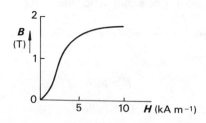

'After magnetic saturation occurs the graph is horizontal' — true or false?

> False. Only the EXTRA flux density stops increasing

The curve rises rapidly at first as a result of a sudden rush of domains becoming aligned with the magnetising field, giving relatively large increases in flux density. As the saturation region is approached, the rate of increase slows down as there are fewer domains remaining to be aligned. After saturation, when all the domains have been aligned, the increase in flux density as H is increased is due solely to the flux produced by the current in the coil. Because domains contain millions of atoms, when they become aligned the increase in flux density is not smooth but takes place in a series of steps, the steepness of the step depending upon how many domains become aligned at the same time. This phenomenon is known as the Barkhausen effect.

Sketch a graph of μ_r as a function of H and comment on its shape. (Use the graph of B against H given in the previous frame and plot $B/\mu_0 H$ to a base of H.)

10

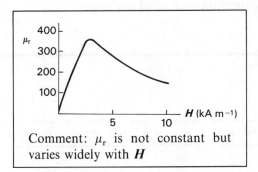

Comment: μ_r is not constant but varies widely with H

Because μ_r varies so widely for different values of H and B it is not sufficient to say that a material has a relative permeability of 100 or whatever. The value of μ_r must be stated together with the value of B to which it is appropriate. The values given in Frame 7 are for arbitrary values of B and are simply to show the wide range of available magnetic materials.

11

HYSTERESIS

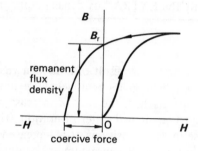

Suppose that we have increased the magnetising field H to the point where the sample has reached saturation. We now consider what happens when we reduce H to zero again. As H is reduced so the flux density B will decrease but it doesn't decrease at the same rate as it increased along the initial magnetisation curve. This is because the aligned domains do not become 'unaligned' at the same rate as they were aligned in the first place. When H reaches zero, some of the domains are still aligned so that there will be a positive value of B. This is called the 'remanent flux density' or the 'residual flux density'. In order to reduce the flux density to zero, it is necessary to apply a negative value of magnetic field strength H. When B has been reduced to zero the corresponding value of H is called the 'coercive force'. These points are all illustrated in the diagram.

What happens if H is further increased negatively?

12

The specimen becomes magnetised again but with the opposite polarity

Again there will be a rapid rise of flux density (but this time of negative polarity) as H is increased negatively, slowing down as saturation is approached. When H is brought back to zero again there will be a residual negative flux density ($-B_r$). This residual flux density has the same magnitude as the positive residual flux density B_r.

What do you suppose happens if we now increase H until it reaches its original saturation value?

13

> The flux density rises again to its original saturation value

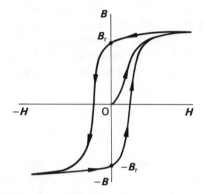

This final rise in B completes what is known as the hysteresis loop (hysteresis means 'lagging behind'; B 'lags behind' H.)

What is the shape of a graph of magnetic flux Φ to a base of magnetising current I? (Think of the relationship between Φ and B, and between I and H.)

14

> The shape is the same as that of the graph of B to a base of H

This is because $B = \Phi/$cross-sectional area (A) of specimen, and $H = NI/l$ where l is the length of the specimen. For a given specimen A, N and l are all constants so that $\Phi \propto B$ and $I \propto H$.

THE MAJOR HYSTERESIS LOOP

The loop obtained when the specimen is taken into saturation is called the major hysteresis loop, and under these conditions the remanent flux density is called the remanence and the coercive force is called the coercivity. There will be an infinite number of smaller loops inside the major loop, each one corresponding to a different limit of H.

According to the hysteresis loop shown above, what is the value of relative permeability μ_r: (a) when $H = 0$, (b) when $B = 0$?

15

$$(a) \ \infty, \quad (b) \ 0$$

In general, the relative permeability at a particular value of B is given by $(1/\mu_0) \times$ the slope of the line joining that point on the B/H curve to the origin, i.e. $\mu_r = B/\mu_0 H$.

This therefore takes no account of the non-linearity of the graph. The *incremental permeability*, μ_i at a particular value of magnetic flux density B is given by the slope of the B/H graph at that point multiplied by $1/\mu_0$, i.e. $\mu_i = dB/\mu_0 dH$.

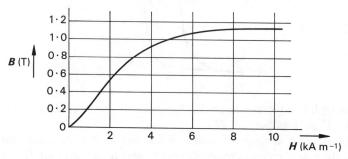

Refer to the curve shown above and determine the relative permeability and the incremental permeability of the sample corresponding to flux densities of (a) 0.5 T and (b) 1.1 T.

16

$$(a) \ \mu_r = 19.8, \ \mu_i = 19.8; \ (b) \ \mu_r = 12.5, \ \mu_i = 3.9$$

HARD AND SOFT MAGNETIC MATERIALS

A material which is easily magnetised, a large value of B resulting from a small value of H, is termed a 'soft' magnetic material. These materials are just as easily demagnetised, only a small coercive force being required to reduce B to zero.

A hard magnetic material, on the other hand, is relatively difficult to magnetise. Once magnetised, however, it requires a large coercive force to reduce the flux density to zero. A hard material therefore has a wide hysteresis loop whereas a soft material has a narrow loop.

17

HYSTERESIS LOSS

When the domains try to align themselves first one way and then the other under the influence of alternating magnetisation, a kind of molecular friction occurs giving rise to an energy loss. This energy loss is called hysteresis loss and it will be shown in Programme 8 to be proportional to the area of the hysteresis loop. In fact, there is an energy loss for every complete cycle of magnetising current measurable in joule per cubic metre of the magnetic material.

If a loop is plotted to a scale of 1 cm = a ampere along the H-axis and 1 cm = b tesla along the B-axis and has an enclosed area of graph paper amounting to S square metre, then the energy loss per cycle will be given by

$$W = S \text{ (cm}^2 \text{ of paper)} \times a \text{ ((A/m)/cm of paper)} \times b \text{ ((T)/cm of paper)}$$

$$= Sab \text{ (T A m}^{-1})$$

Now $(\text{T A m}^{-1}) = (\text{Wb m}^{-2} \text{ A m}^{-1}) = (\text{V s m}^{-2} \text{ A m}^{-1}) = (\text{V A s m}^{-3})$ $= (\text{W s m}^{-3}) = (\text{J m}^{-3})$. This means that the product (BH) has the unit of energy per unit volume.

$$\therefore \text{ energy loss per cycle} = Sab \text{ joule per cubic metre of material.}$$

If the specimen has a volume v, energy loss = $Sabv$ joule per cycle.

What will be the total power loss (in watts) if the specimen is taken through alternating magnetisation at a rate of f cycles per second?

18

$$\boxed{P = Sabvf \qquad \text{(W)}}$$

Joules per cycle multiplied by cycles per second gives joules per second and the rate of doing work is power measured in watts.

Example

The hysteresis loop for a sample of magnetic material is drawn using scales of 1 cm to 100 A m^{-1} and 1 cm to 0.1 T. The area of the loop is 54 cm^2. If the volume of the specimen is 5000 cm^3, calculate the hysteresis loss at a frequency of 50 Hz.

The solution is given in the next frame. Try it yourself, first.

19

$$\boxed{135 \text{ W}}$$

This is obtained using the formula $P = Sabfv$ (W) in which $S = 54 \text{ cm}^2$, $a = (100 \text{ A m}^{-1})/\text{cm}$, $b = (0.1 \text{ T})/\text{cm}$, $f = 50 \text{ Hz}$, $v = 5000 \times 10^{-6} \text{ m}^3$.

Now decide which of the following statements are true:

(a) Diamagnetic materials align themselves with their longer axis at right angles to the direction of a non-uniform magnetic field.

(b) Paramagnetic materials are less permeable than a vacuum.

(c) Ferromagnetism is a sub-group of paramagnetism.

(d) Diamagnetic materials have permeabilities in the range 0.5–0.9.

(e) Ferromagnetic materials can have μ_r of the order of 1000.

(f) The magnetic field of a current-carrying coil is greatly increased if an iron core is inserted into it.

(g) A domain contains about 10 000 atoms.

(h) In general, $\boldsymbol{B} = \mu_0 \mu_r \boldsymbol{H}$.

(i) The unit of permeability is the tesla per ampere metre.

(j) The graph of $\boldsymbol{B}$ against $\boldsymbol{H}$ increases until saturation, after which it becomes horizontal.

(k) μ_r is constant for a given material.

(l) For a given sample of magnetic material the shape of the graph of Φ against I is similar to that of $\boldsymbol{B}$ against $\boldsymbol{H}$.

(m) A material having a high coercivity is required for use in electromagnetic relays.

(n) The intercept of the major loop on the $\boldsymbol{B}$-axis is termed the remanence.

20

$$\boxed{\text{True: a, c, e, f, h, l, n}}$$

(b) They are more permeable. (Frame 4)

(d) Most diamagnetic materials have $\mu_r > 0.999$. (Frame 7)

(g) Domains contain many millions of atoms. (Frame 5)

(i) The unit of μ is the unit of $\boldsymbol{B}$/the unit of $\boldsymbol{H}$ (T m A^{-1}). (Frame 7)

(j) After saturation it rises with a slope of μ_0. (Frame 9)

(k) μ_r varies widely with variations in magnetic flux density. (Frame 10)

(m) Electromagnets need to be easily demagnetised. (Frame 5)

21

SUMMARY OF FRAMES 1–20

Diamagnetic materials are marginally less permeable than a vacuum.
Paramagnetic materials are marginally more permeable than a vacuum.
Ferromagnetic materials are many times more permeable than a vacuum.
   The process of magnetising a specimen of iron is one of aligning its domains in the direction of the magnetising field. When they have all been aligned, a state of magnetic saturation is said to have been reached.
The graph of B against H is called a magnetisation or saturation curve.
For alternating magnetisation, the graph of B against H is called a hysteresis loop.
A hysteresis loop taken into saturation is called a major loop.
The intercept of a major loop on the B-axis is termed remanence.
The intercept of a major loop on the H-axis is termed coercivity.
A 'hard' material is difficult to magnetise and has a high coercivity.
A 'soft' material is easy to magnetise and has a low coercivity.
There is an energy loss associated with hysteresis which is proportional to the area enclosed by the hysteresis loop.
Relative permeability $\quad \mu_r = B/\mu_0 H$.
Incremental permeability $\mu_i = dB/\mu_0 \, dH$.
Absolute permeability $\quad \mu = \mu_0 \mu_r$.

22

EXERCISES

(i) The hysteresis loop for a sample of magnetic material is drawn to the following scales: H-axis, 1 cm = 150 A/m; B-axis, 1 cm = 0.2 T. The area of the loop is 40 cm^2. Calculate the hysteresis loss in watt per kilogram, assuming the density of the material to be 8000 kg/m^3 and the frequency is 50 Hz.

(ii) Half of a hysteresis loop is given by the following points:

B (T):	0.1	0.25	0.5	0.6	0.75	0.7	0.6	0.54	0.34	0
H (A m^{-1}):	175	200	260	300	400	230	60	0	−120	−170

Plot the loop and from it determine:

(a) the hysteresis loss if the volume of the specimen is 0.006 cm^3,

fray?

(b) the coercivity,
(c) the remanence,
(d) the relative permeability when $B = 0.55$ T (rising),
(e) the incremental permeability when $B = 0.4$ T (rising).

23

MAGNETIC CIRCUITS

An assembly of resistors which form a closed path around which an electric current flows is called an electric circuit. The current in the circuit is produced by a source known as an electro-motive force (e.m.f.), symbol E. Similarly, an assembly of magnetic materials which form a closed path around which a magnetic flux is established is called a magnetic circuit.

In the magnetic circuit, the source of the flux is called an m.m.f. (magneto-motive force) and is usually a current-carrying coil. The symbol for m.m.f. is F and its unit is the ampere. The circuits themselves usually consist of more of less complicated shapes of iron or some other ferromagnetic material. Some have no air gaps in them, some have air gaps designed into them and others have unwanted air gaps in them.

Give an example of a circuit having an air gap.

24

An air gap is essential in the magnetic circuit of an electric motor in order that motion can take place.

For the purpose of calculation, magnetic circuits are divided into sections, each of which has a uniform cross-sectional area throughout its length. In subdividing circuits in this way, any air gap must of course be regarded as being a section of the circuit.

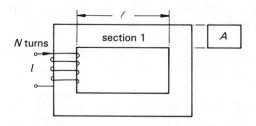

The diagram shows a magnetic circuit consisting of an iron core and a single coil of N turns carrying a current of I ampere.

Give an expression for the magneto-motive force (m.m.f.) in this circuit.

25

$$\boxed{F = NI \qquad (A)}$$

RELUCTANCE

Let us now consider section 1 of the circuit given in the previous frame. It is of iron, has a uniform cross-sectional area over its length, l, and the flux of $\boldsymbol{B}$ in it is Φ, all of which is contained within its volume.

The flux density at all points in this circuit element is therefore given by $\boldsymbol{B} = \Phi/A$ and the field strength $\boldsymbol{H} = \boldsymbol{B}/\mu = \Phi/\mu A$ (A m^{-1}).

The magnitude of the magnetic potential difference between the ends of the section is given by $\int_L H \cos\theta \, dl$ (A) (see Programme 5, Frame 19). Since $\boldsymbol{B}$ is constant, so too is $\boldsymbol{H}$. Also $\cos\theta = 1$ since $\boldsymbol{H}$ is acting in the direction of the path L so the potential difference is

$$H \int_L dl = Hl$$

Since $\boldsymbol{H} = \Phi/\mu A$, then $Hl = \Phi l/\mu A$ (A).

$l/\mu A$ is called the reluctance of the magnetic circuit element and is a measure of how difficult it is to establish magnetic flux in the element. The symbol for reluctance is S and its unit is the reciprocal Henry (H^{-1}).

What is the electric circuit analogue of reluctance?

26

$$\boxed{\text{Resistance } (R = l/\sigma \, A)}$$

Note that both resistance and reluctance increase as the length of the circuit element increases, and both increase as the cross-sectional area of the circuit element decreases. The circuit constant, σ (conductivity), is the electrical circuit analogue of μ (permeability).

A piece of magnetic material measures 20 cm $\times$ 30 cm in cross-section and is 60 cm long. Assuming that $\mu_r = 660$, calculate its reluctance.

27

$$\boxed{12\,057 \text{ H}^{-1}}$$

... because $S = l/\mu A$ where $l = 0.6$ m, $A = 0.2 \times 0.3$ m^2, $\mu = 660 \times 4\pi \times 10^{-7}$ H m^{-1}.

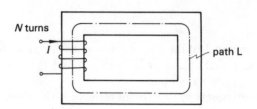

Let us now apply Ampere's circuital law to the path L in the magnetic circuit shown. As a first consideration, we shall assume:

(a) all sections of the circuit have the same cross-sectional area A and that they have a total length l;
(b) all of the magnetic flux is retained within the circuit;
(c) the magnetic field strength is constant throughout the circuit and directed in the direction of the path L at all points.

Under these conditions, applying Ampere's circuital law gives

$$H \int_L dl = \sum I \Rightarrow Hl = NI \qquad \text{(A)}$$

As we saw in Frame 25 $Hl = \Phi l/\mu A = \Phi S$
Also $\qquad\qquad\qquad NI = F$
so that $\qquad\qquad\quad F = \Phi S$
i.e. $\qquad\qquad$ (m.m.f. = flux $\times$ reluctance)
What is the equivalent equation in electric circuit theory?

28

$$\boxed{E = IR \qquad \text{(V)}}$$

Until you become expert at dealing with magnetic circuits it is useful to keep in mind the electric circuit analogies. Draw up a list of analogues.

Electric circuit	Magnetic circuit
e.m.f. E	m.m.f. F
current I	flux Φ
resistance R	reluctance S
conductivity σ	permeability μ
$E = IR$	$F = \Phi S$
$R = l/\sigma A$	$S = l/\mu A$

The reciprocal of reluctance is *permeance*, symbol Λ so that $\Lambda = \mu A/l$ (H).

THE PERMEANCE OF AN IRON RING

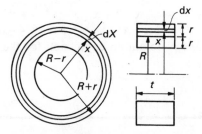

The iron ring shown has a rectangular cross-section ($2r \times t$) and may be considered to be made up of a series of concentric cylinders. One of these, having a radius x and a radial thickness dx, is shown. The mean radius of the ring is R, the inner radius is $(R - r)$ and the outer radius is $(R + r)$. If the elementary cylinder is taken to be small enough, the flux density may be considered to be uniform over its cross-sectional area ($t \times dx$).

Using the above formula, we have for the permeance of the thin cylinder

$$d\Lambda = \frac{\mu t \, dx}{2\pi x} \quad \text{(H)} \qquad \text{where } t \, dx = A \text{ and } 2\pi x = l$$

To obtain the permeance, Λ, of the entire ring we add the permeances of all the thin cylinders which make up the ring.

i.e.
$$\Lambda = \Sigma \, d\Lambda = \int_{R-r}^{R+r} (\mu t/2\pi x) \, dx \quad \text{(H)}$$

Now you complete the problem to obtain an expression for the permeance, Λ.

30

$$\Lambda = \frac{\mu t}{2\pi} \ln\left[\frac{R+r}{R-r}\right] \qquad \text{(H)}$$

Since μ, t and 2π are constants, they are taken through the integral sign and we are

left with

$$\Lambda = \frac{\mu t}{2\pi} \int_{R-r}^{R+r} \frac{\mathrm{d}x}{x} = \frac{\mu t}{2\pi} \ln[x]_{R-r}^{R+r} \qquad \text{(H)}$$

$$= (\mu t/2\pi)[\ln(R+r) - \ln(R-r)] \qquad \text{(i)}$$

which is equivalent to the answer in the box.

Now from equation (i) we have $\Lambda = (\mu t/2\pi)[\ln(1+r/R) - \ln(1-r/R)]$.

Using the expansions

$$\ln(1+x) = x - x^2/2 + x^3/3 - x^4/4 + x^5/5 - \dots$$
$$\text{and} \quad \ln(1-x) = -x - x^2/2 - x^3/3 - x^4/4 - x^5/5 - \dots$$

and remembering that $R \gg r$, obtain an expression for the approximate permeance of the ring.

31

$$\Lambda = \frac{\mu(t \times 2r)}{2\pi R} \qquad \text{(H)}$$

After expanding and collecting together terms, you should have obtained

$$\Lambda = \frac{\mu t}{2\pi}\left[\frac{2r}{R} + \frac{2}{3}\left(\frac{r}{R}\right)^3 + \frac{2}{5}\left(\frac{r}{R}\right)^5 + \dots\right]$$

which reduces to $\dfrac{\mu t}{2\pi}\dfrac{2r}{R}$, ignoring all other terms since $R \gg r$. This is equal to

$$\mu \times \frac{\text{the cross sectional area of the ring (call it } A)}{\text{the length of the geometric mean line of the ring } (l)}.$$

Give an approximate expression for the reluctance of the ring.

32

$$S = 1/\Lambda = l/\mu A \qquad \text{(H)}$$

Since $\qquad\qquad\qquad F = \Phi S = \Phi/\Lambda$

then $\qquad\qquad\qquad\quad \Phi = F\Lambda$

and the flux in the elementary thin cylinder of the ring in Frame 29 is thus

$$d\Phi = Fd\Lambda = NI\frac{\mu t\,dx}{2\pi x} \qquad \text{(Wb)}$$

The flux density in the thin cylinder i.e. at radius x, is therefore given by

$$B = \frac{d\Phi}{\text{area of thin cylinder}} = \frac{d\Phi}{t\,dx} = \frac{NI\mu t\,dx}{2\pi x \times t\,dx} \qquad \text{(T)}$$

$$= \frac{NI\mu}{2\pi x} \qquad \text{(T)}$$

This will vary from a maximum of $\dfrac{NI\mu}{2\pi(R-r)}$ at the inner radius of the ring to a

minimum of $\dfrac{NI\mu}{2\pi(R+r)}$ at the outer radius of the ring.

Is this a large variation?

33

$$\boxed{\text{No — so long as } R \gg r}$$

Under these conditions the flux density can be regarded as being substantially constant over the cross-section of the ring and is usually taken to be the value along the geometric mean line. Because the magnetic circuits which are met in practice are often rather complicated in geometry, very accurate calculation of permeance and consequently of the flux–m.m.f. relationship can be most complicated. In such cases we make use of the approximate method, using the geometric mean line (which may be thought of as the mean path of the flux) for the purposes of calculation. Fortunately great accuracy is rarely required and the approximate method is acceptable in most cases.

34

Example

A mild steel ring having an inside diameter of 14 cm and an outside diameter of 18 cm is of rectangular cross-section with a thickness of 1.5 cm. Calculate, assuming a relative permeability of 1180:

(a) the exact value of the reluctance of the ring,
(b) the permeance of the ring using the approximate (mean flux path) method,
(c) the current required in a coil of 80 turns to produce a flux of 350 μWb.

Solution

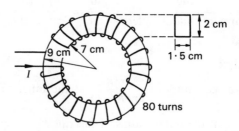

The ring is uniformly wound so that the flux density is uniform.
All of the flux produced is assumed to be contained within the ring.

(a) The 'exact' value of permeance is given by $\dfrac{\mu t}{2\pi}\ln\left[\dfrac{R+r}{R-r}\right]$ (H).

Putting in the values $\mu = 1180 \times 4\pi \times 10^{-7}$, $t = 0.015$ m, $(R+r) = 0.9$ m and $(R-r) = 0.7$ m, we obtain $\Lambda = 0.8896 \times 10^{-6}$ H.

The reluctance $S = 1/\Lambda = \underline{1.124 \times 10^6 \text{ H}^{-1}}$.

(b) Using the approximate method, the permeance of the ring is given by

$$\Lambda = \frac{\mu A}{2\pi R} = \frac{1180 \times 4\pi \times 10^{-7} \times 0.02 \times 0.015}{2\pi \times 0.08} = \underline{0.885 \ \mu\text{H}}$$

Comparing this with the value obtained in part (a), we see that the error involved is only $-\dfrac{46}{8896} \times 100\% = -0.52\%$.

(c) $F = S\Phi = 1.124 \times 10^{-6} \times 350 \times 10^{-6}$ A $= 393.4$ A.

$\therefore \quad NI \ (=F) = 393.4$ A

and $I = 393.4/80 = \underline{4.92 \text{ A}}$

SERIES MAGNETIC CIRCUITS

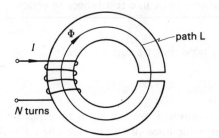

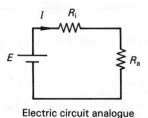

Electric circuit analogue

The diagram shows an iron ring with a radial air gap cut in it. The electric circuit analogue is also shown. The m.m.f. is a coil of N turns carrying I ampere and this produces Φ weber in the ring. We will assume that:

(a) all of the flux is retained within the circuit so that the flux in the iron element (which has a reluctance S_i) is the same as that in the air gap element (which has a reluctance S_a);

(b) the flux passes straight across the air gap so that its effective cross-sectional area is the same as that of the iron—it follows that the flux density in the iron is the same as that in the air gap;

(c) the magnetic field strength is constant at H_i in the iron over its length l_i, and is constant at H_a in the air gap over its length l_a.

Now apply Ampere's circuital law to the path L.

36

$$\boxed{H_i l_i + H_a l_a = NI}$$

$$\oint_L H \cos \theta \, dl = H_i l_i + H_a l_a \qquad \text{and the path links } NI \text{ ampere.}$$

Now we have seen (Frame 27) that $Hl = S\Phi$

$$\therefore \qquad H_i l_i + H_a l_a = S_i \Phi_i + S_a \Phi_a$$

Also $NI = F$

$$\therefore \qquad F = S_i \Phi_i + S_a \Phi_a = \Phi(S_i + S_a) \qquad \text{since } \Phi_i = \Phi_a$$

Compare this with the equation of the analogous electric circuit:

$$E = I(R_i + R_a)$$

What is the equivalent reluctance of a number of reluctances in series?

> The sum of the individual reluctances—just like resistances in series

In solving problems on magnetic circuits, note that we can use either

$$F = S\Phi \qquad \text{or} \qquad F = Hl$$

Which we use depends on the given information. If we are given information from which we are easily able to calculate the reluctance of the circuit, then $F = S\Phi$ is probably more convenient. If, however, we are given the relationship between H and B in the form of a graph, we use $F = Hl$.

Example

A steel ring has a mean radius of 10 cm and a uniform cross-sectional area of 5 cm². A radial air gap of length 1.5 mm is cut in the ring. If the relative permeability of the steel is constant at 1250, calculate the m.m.f. of a coil required to produce a magnetic flux of 500 μWb in the air gap. State any assumptions made.

Solution

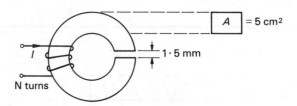

We assume that all of the flux produced by the coil is retained within the magnetic circuit so that the flux in the steel and in the air gap is the same.

The reluctance of the ring, $S_i = \dfrac{l_i}{\mu A_i} = \dfrac{2\pi \times 0.1}{1250 \times 4\pi \times 10^{-7} \times 5 \times 10^{-4}} = 0.8 \times 10^6 \text{ H}^{-1}$

Note that the length of the iron is taken to be the mean circumference of the ring. It is not reduced by the length of the air gap because that level of accuracy is not appropriate in these problems.

The flux in the ring, $\Phi = 500 \times 10^{-6}$ Wb so that $S_i\Phi = 500 \times 0.8 = 400$ A.

The next step is to calculate the reluctance of the air gap, assuming that its cross-sectional area is the same as that of the ring (i.e. that the flux passes directly across it). What do you make it?

$$\boxed{2.39 \times 10^6 \text{ H}^{-1}}$$

The reluctance of the air gap, $S_a = \dfrac{l_a}{\mu_0 A_a} = \dfrac{1.5 \times 10^{-3}}{4\pi \times 10^{-7} \times 5 \times 10^{-4}} = 2.39 \times 10^6 \text{ H}^{-1}$.

The flux $\Phi = 500 \ \mu$Wb so that $S_a\Phi = 2.39 \times 500 = 1195$ A.

The total m.m.f., $F = S_i\Phi + S_a = 400 + 1195 = \underline{1595 \text{ A}}$

Note that the air gap of only 1.5 mm requires an m.m.f. almost three times as great as the iron path which is more than 600 mm in length.

Example

A magnetic circuit consists of two sections in series. Section 1 is of mild steel, having a length of 50 cm and a uniform cross-sectional area of 3 cm². Section 2 is an air gap of length 1 mm and has the same cross-sectional area as the steel.

Calculate the current required in a coil of 150 turns wound on the steel in order to produce a magnetic flux of 400 μWb in the circuit, assuming that the flux is the same in both sections of the circuit.

The magnetic characteristic of the mild steel is given below:

B (T):	0.8	1.1	1.35	1.5	1.58	1.62	1.68
H (A m^{-1}):	250	500	1000	2000	3000	4000	5000

Solution

Since we are given the relationship between B and H, it is more convenient in this problem to use $F = Hl$.

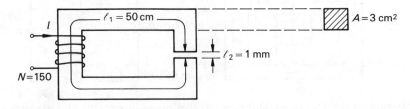

The total m.m.f. is given by $F = (Hl)_1 + (Hl)_2$.

Now determine the flux density B_1 in section 1 and by drawing the graph of B against H from the data in the table above, obtain the corresponding value of magnetic field strength H_1. You must draw the graph. Do not try to guess a value for H by interpolation from the table.

39

$$\boxed{B_1 = 1.34 \text{ T}; \ H_1 = 900 \text{ A m}^{-1}}$$

$$B_1 = \frac{\text{flux in the ring}}{\text{cross sectional area of the ring}} = \frac{400 \times 10^{-6}}{3 \times 10^{-4}} = 1.33 \text{ T}$$

From your graph you should find that the corresponding value for H_1 is 900 A m^{-1}.

$$\therefore \quad H_1 l_1 = 900 \times 0.5 = 450 \text{ A}$$

Considering section 2 (an air path so that no graph is required):
Magnetic flux density, $B_2 = 1.33$ T (since the flux and the cross-sectional area are the same as for section 1).
Magnetic field strength, $H_2 = B_2/\mu_0 = 1.058 \times 10^6$ A m^{-1}.

$$\therefore \quad H_2 l_2 = 1.058 \times 10^6 \times 10^{-3} = 1058 \text{ A}$$

The total m.m.f., $F = 1058 + 450 = 1508$ A and this equals NI. The current in a coil of 150 turns is thus $1508/150 = \underline{10.05 \text{ A}}$.

40

PARALLEL MAGNETIC CIRCUITS

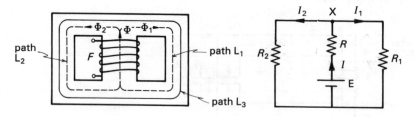

The diagram on the left above shows a parallel magnetic circuit and alongside it is the electrical circuit analogy. All of the flux produced appears in the centre limb where the m.m.f. is located, just as, in the electric circuit, all of the current flows in the central branch where the e.m.f. is located.

What determines how much of the flux appears in the left-hand limb of the magnetic circuit?

41

> The reluctances of the left- and right-hand limbs of the circuit.
> The greater amount of flux appears where the reluctance is lower.

... just as the greater current will flow in the branch having the lower resistance in the electrical circuit.

Concentrating for a moment on the electrical circuit, we can write down three Kirchhoff's voltage law equations:

$$E = IR + I_2 R_2 \qquad \text{for the left-hand mesh}$$
$$E = IR + I_1 R_1 \qquad \text{for the right-hand mesh}$$
$$0 = I_2 R_2 - I_1 R_1 \qquad \text{for the outer mesh}$$

The third one could of course be obtained from the other two.

Kirchhoff's current law applied to the node X gives

$$I = I_1 + I_2$$

What are the analogous equations for the magnetic circuit?

42

> $$F = \Phi S + \Phi_1 S_1 \qquad \text{for the path } L_1$$
> $$F = \Phi S + \Phi_2 S_2 \qquad \text{for the path } L_2$$
> $$\Phi_1 S_1 = \Phi_2 S_2 \qquad \text{for the path } L_3$$
> $$\Phi = \Phi_1 + \Phi_2$$

In each of the magnetic potential equations we may replace any ΦS by the equivalent Hl so we obtain:

$$F = Hl + H_1 l_1$$
$$F = Hl + H_2 l_2$$
$$\text{and} \quad H_1 l_1 = H_2 l_2$$

where l, l_1 and l_2 are lengths of the centre, right-hand and left-hand limbs respectively.

Armed with these three equations together with the four in the box, you should be able to solve any parallel magnetic circuit problem you are likely to meet.

43

Example

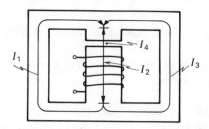

The magnetic circuit shown has the following dimensions:

left-hand limb: path l_1, length 34 cm, cross-sectional area 5 cm^2;
centre limb: path l_2, length 15 cm, cross-sectional area 8 cm^2;
right-hand limb: path l_3, length 34 cm, cross-sectional area 5 cm^2;
air-gap: path l_4, length 1 mm, cross-sectional area 8 cm^2.

The limbs are of mild steel whose magnetic characteristic is given in Frame 38. Calculate the m.m.f. required of the coil wound on the centre limb to produce a flux of 1 mWb in the air gap.

Solution

Applying the magnetic circuit potential equation to the closed path which consists of l_1, l_2 and l_4, we have $F = H_1 l_1 + H_2 l_2 + H_4 l_4$ where H_1, H_2 and H_4 are the field strengths acting over the sections l_1, l_2 and l_4 respectively.

Considering section l_4, the magnetic flux is $\Phi_4 = 1 \times 10^{-3}$ Wb
thus the flux density is $B_4 = 10^{-3}/(8 \times 10^{-4}) = 1.25$ T
the magnetic field strength $H_4 = B_4/\mu_0 = 995$ kA m^{-1}
and the magnetic 'potential drop' is $H_4 l_4 = 995 \times 10^3 \times 10^{-3} = 995$ A.

In considering section l_2 we assume that the flux in the centre limb is the same as that in the air gap. It follows that the flux densities will also be the same because the cross-sectional areas are the same.
The flux density is thus $B_2 = 1.25$ T.
From the graph in Frame 38, the corresponding H_2 is 750 A m^{-1}.
The magnetic potential drop is $H_2 l_2 = 750 \times 0.15 = 112.5$ A.

In order to calculate the potential drop in the left-hand limb, we need first to determine the flux, Φ_1, in it. Do that now.

$$\boxed{\Phi_1 = 0.5 \text{ mWb}}$$

The reluctances of the two outer limbs are the same so that the total flux, Φ_4 will divide equally between the two. It follows that $\Phi_1 = \Phi_3 = 0.5$ mWb. (Check that the equation $\Phi_4 = \Phi_1 + \Phi_3$ is satisfied.)
The flux density is $\qquad B_1 = 0.5 \times 10^{-3}/5 \times 10^{-4} = 1$ T.
From the graph $H_1 = 400$ A m^{-1}.
The magnetic potential drop is $H_1 l_1 = 400 \times 0.34 = 136$ A.
The m.m.f. is equal to the sum of the potential drops around the circuit so we have
$F = 995 + 112.5 + 136 = \underline{1243 \text{ A}}$.

45

MAGNETIC LEAKAGE

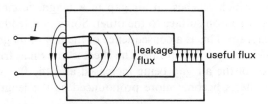

So far we have assumed that all of the flux produced by the m.m.f. in a magnetic circuit is confined to the paths which comprise the circuit. In practice this is unlikely to be the case. In an electrical circuit the copper conductors are of the order of 10^{29} times more attractive than air to electric current which consequently confines itself to the circuit. In a magnetic circuit the ferromagnetic paths are only of the order of 10^3 times more attractive than air to magnetic flux and it is therefore virtually impossible to retain all of it within the circuit. Some of it will pass out through the sides at some points and re-enter through the sides at other points. In so doing it might miss out the part of the circuit where it is needed and is therefore useless. This useless flux is called leakage flux and the remainder is called useful flux.

The ratio $\dfrac{\text{total flux produced}}{\text{useful flux}}$ is called the leakage factor.

A magnetic circuit requires a flux of 200 mWb in an air gap and this is produced by a coil wound on an iron limb 50 cm $\times$ 40 cm in cross-section. If the flux density in the iron limb is 1.2 T, what is the leakage factor?

$$\boxed{1.2}$$

Clearly, the element of the circuit which has the coil wound on it will have the greatest flux because all of the lines of force must link one or more turns of this 'exciting' coil as it is called.

In this case, the total flux $= 1.2 \times 0.5 \times 0.4 = 0.24$ Wb ($\mathbf{B} \times$ area).
The useful flux $= 0.2$ Wb.
The leakage factor $= 0.24/0.2 = 1.2$.

FRINGING FLUX

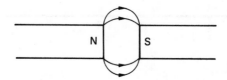

Not all of the flux which reaches an air gap in a magnetic circuit passes directly across the gap from one iron surface to the other. Some of it tends to bulge outwards as shown in the diagram. This is a consequence of the law of magnetic flux refraction which will be considered in a later frame. The effect is known as fringing and it results in the effective area of the air gap being increased and the flux density there being reduced. Fringing effects become more pronounced as the length of the air gap is increased.

Example

A magnetic circuit consists of an iron core of cross-sectional area 4 cm^2 in which is cut an air gap where a flux density of 0.9 T is required. There is a leakage factor of 1.2 and fringing increases the cross-sectional area of the gap by 10%. What must be the flux density in the iron?

Solution

Area of the gap $= 1.1 \times$ that of the iron $= 1.1 \times 4 \times 10^{-4} = 4.4 \times 10^{-4}$ m^2.
Flux required in the gap is therefore $0.9 \times 4.4 \times 10^{-4} = 396$ μWb.
There is a leakage factor of 1.2 so that the flux in the iron must be

$$1.2 \times 396 \ \mu\text{Wb} = 475 \ \mu\text{Wb}$$

The flux density in the iron is therefore $475 \times 10^{-6}/4 \times 10^{-4} = \underline{1.19 \text{ T}}$.

STATIONARY MAGNETIC FIELDS 2

47

Which of the following statements are true?

(a) In magnetic circuit calculations, unwanted air gaps are ignored but air gaps designed into the circuit are counted as circuit elements.
(b) Magnetomotive force is given by $F = NI/l$.
(c) The unit of magneto-motive force is the ampere.
(d) The reluctance of a magnetic circuit element increases with its length.
(e) The permeance of a magnetic circuit element increases with its area.
(f) The unit of reluctance is the Henry.
(g) The permeance of a magnetic circuit element is a measure of the difficulty in setting up a magnetic flux in it.
(h) For any magnetic circuit element $S\Phi = Hl$.
(i) The total reluctance of a number of magnetic circuit elements in series is the sum of the reluctances of the individual circuit elements.
(j) The electrical circuit analogy of permeance is conductance.
(k) The permeance of an iron ring is given approximately by $\mu A/l$ where A is its cross-sectional area and l is its geometric mean length.
(l) A very small air gap whose length is very much less than that of the iron parts of the circuit has a negligible effect on the circuit operation.
(m) A typical value for the leakage factor in a magnetic circuit is 0.9.
(n) Fringing flux is another name for leakage flux.
(o) Fringing effectively increases the flux density in the air gap.

48

| True: c, d, e, h, i, j, k |

(a) All air gaps must be considered as a part of the circuit.   (Frame 24)
(b) m.m.f. is measured in ampere and is NI.   (Frame 23)
(f) The unit of reluctance is the reciprocal Henry.   (Frame 25)
(g) The higher the permeance, the easier it is to establish flux.   (Frame 29)
(l) A small air gap has a very significant effect.   (Frame 38)
(m) Leakage factor is always bigger than 1.   (Frame 45)
(n) Fringing occurs only at air gaps.   (Frame 46)
(o) Fringing effectively decreases the flux density in the gap.   (Frame 46)

49

SUMMARY OF FRAMES 23–48

New quantities and units

Quantity	Symbol	Unit	Abbreviation
Magnetomotive force, m.m.f.	F	Ampere	A
Reluctance	S	Reciprocal Henry	H^{-1}
Permeance	Λ	Henry	H

Formulae

$S = \dfrac{l}{\mu A}$ (Henry)$^{-1}$ (the reluctance of a circuit element of length l, and cross sectional area A)

$\Lambda = 1/S$ Henry (permeance is the reciprocal of reluctance)

$F(=NI) = \boldsymbol{H}l = S\Phi$ (magnetic potential relationships)

Leakage factor $= \dfrac{\text{total flux}}{\text{useful flux}}$ which is always > 1

50

EXERCISES

(i) Calculate the reluctance of an iron bar 12 cm long and having a cross-sectional area of 4 cm^2 if the relative permeability is 900.

(ii) A series magnetic circuit consists of a steel ring having a mean diameter of 20 cm and a cross-sectional area of 6 cm^2. Calculate the m.m.f. required to produce a flux of 0.6 mWb in a radial air gap 2 mm long cut into the ring. Neglect leakage and fringing. The steel has a magnetic characteristic as follows:

$$
\begin{array}{lccccccc}
\boldsymbol{B}\,(\text{T}): & 0.1 & 0.3 & 0.5 & 0.7 & 0.9 & 1.1 & 1.3 \\
H\,(\text{A m}^{-1}): & 140 & 330 & 480 & 650 & 860 & 1200 & 1700
\end{array}
$$

(iii) Three steel bars are effectively in parallel. The two outer bars are each 50 cm long and 20 cm^2 in cross-section. The centre bar is 20 cm long and 8 cm^2 in cross-section. Calculate the current required in a coil of 150 turns on the centre limb to produce a flux of 0.5 mWb there. The steel has the same magnetic characteristic as that in (ii) above.

51

PERMANENT MAGNETS

Permanent magnets play an important part in many applications (for example, in meters and loudspeakers) and, as we have seen, they are made from magnetically hard materials. Having been magnetised, which requires a large value of H, and then the magnetising field having been removed, the permanent magnet finds itself at the point $(0, B_r)$ on the hysteresis loop. As we saw in Frame 14 this represents the remanence. From this point the loop moves into the second quadrant where B is positive and H is negative and it is this quadrant which is of particular interest in permanent magnet calculations.

Most permanent magnet applications include an air gap and it is important to investigate the nature of the B-field and the H-field in such cases.

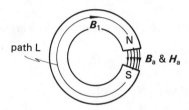

The diagram shows a C-shaped permanent magnet. The direction of the B-field and the H-field between the poles is as shown, the lines of force leaving a North pole and entering a South pole.

Now apply Ampere's circuital law to the path L and hence determine the direction of the H-field in the iron.

52

> The H-field direction is anticlockwise in the iron

Applying Ampere's law to path L yields $H_i l_i + H_a l_a = 0$ since there is no current linked with the path. It follows that $H_i l_i = -H_a l_a$ and because l_i and l_a are both positive, H_i must be of opposite sign to H_a.

In the second quadrant of the hysteresis loop, B and H are of opposite sign and so the flux density in the iron is acting in a clockwise direction.

If the cross-sectional area of the air gap is the same as that of the iron, the magnitude of the flux density will be the same in both.

What will be the relative magnitudes of H_a and H_i?

53

Since the air gap length is always very much smaller than the iron path, it follows that $H_a \gg H_i$.

TO FIND THE OPERATING POINT FOR A PERMANENT MAGNET

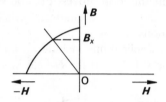

If the required flux density is B_x, then assuming no fringing effects

$$B_i = B_a = B_x$$

Now $B_a = \mu_0 H_a$ so that $B_i = \mu_0 H_a$ (T).
But we have seen that $H_i l_i = -H_a l_a$ (A).

$$\therefore \qquad H_a = -\frac{H_i l_i}{l_a} \quad (\text{A m}^{-1}) \qquad \text{and} \qquad B_i = -\mu_0 \times \frac{l_i}{l_a} H_i \quad (\text{T})$$

... which is the equation of a straight line through the origin and having a slope of $-\mu_0(l_i/l_a)$ as shown in the diagram.

Example

A permanent magnet is 15 cm long and 2 cm^2 in cross-section. Neglecting leakage and fringing, calculate the flux density in an air gap 0.6 cm long. The demagnetisation curve for the material of the magnet is as follows:

B (T):	0	0.25	0.5	0.7	0.8	0.95
H (kA m^{-1}):	-100	-80	-60	-40	-20	0

Solution

The slope of the load line is $-\mu_0 \dfrac{l_i}{l_a} = -4\pi \times 10^{-7} \dfrac{0.15}{0.006} = -3.142 \times 10^{-5}$ H m^{-1}.

Now complete the problem by plotting the curve and the load line to find B_x.

54

Flux density in the air gap $= 0.8$ T

Here is the working:

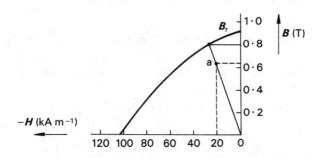

First we plot the demagnetisation curve. Then, to draw the load line, we need one point and the origin. For any value of H, we can find the corresponding value of B by multiplying by the slope (-3.14×10^{-5}). If we take $H = -20$ kA m^{-1} then $B = -3.14 \times 10^{-5} \times (-20) \times 10^3 = 0.628$ T. This fixes point a and the load line is drawn through it and the origin.

The load line intercepts the demagnetisation curve at B_x and, assuming no fringing this is also the flux density in the air gap.

What would be the effect of bridging the air gap by means of a piece of soft iron?

55

The flux density would rise to B_r

This is because, with the air gap removed, $H_a l_a = 0$ and so $H_i l_i = 0$ in accordance with $\oint_L H \, dl = 0$ (Ampere's law applied to this circuit). Since $l_i \neq 0$, then $H_i = 0$, and the corresponding value of B_i is B_r.

The presence of the air gap reduces the flux density of a permanent magnet and to retain its magnetism the air gaps are removed using these pieces of soft iron which are called 'keepers'. In practice, the flux density might not be fully restored to the original B_r value, especially if the magnetic has been in use for a very long time.

56

THE MINIMUM VOLUME REQUIRED FOR A PERMANENT MAGNET

Good quality permanent magnets are made from materials which are very expensive (they can cost thousands of pounds per kilogram!). It is very important therefore to obtain the required amount of magnetic flux in the air gap with as little magnetic material as possible.

We have that, in magnitude,

$$H_i l_i = H_a l_a \quad \text{(Frame 52) so that} \quad l_i = \frac{H_a l_a}{H_i} \text{(m)} \tag{i}$$

Also, neglecting magnetic leakage, the flux in the air gap is the same as that in the iron part of the circuit, i.e. $\Phi_i = \Phi_a$.

If the area of the magnet is A_i and that of the air gap is A_a,

then
$$B_i A_i = B_a A_a \quad \text{and} \quad A_i = \frac{B_a A_a}{B_i} (\text{m}^2) \tag{ii}$$

Now the volume of the magnet is $A_i l_i$, so multiplying (i) by (ii):

$$A_i l_i = \frac{H_a l_a}{H_i} \frac{B_a A_a}{B_i} \quad (\text{m}^3)$$

Substituting for $H_a = B_a/\mu_0$, we have for the volume V of the magnet:

$$V = \frac{B_a^2 l_a A_a}{\mu_0(H_i B_i)} \quad (\text{m}^3)$$

For a given air gap and a given value of air gap flux, therefore, what is the condition that the volume of the magnet material shall be a minimum?

57

The product $H_i B_i$ must be as large as possible

The product $H_i B_i$ is measured in joule per cubic metre (see Frame 17) and for any point on the demagnetisation curve it is proportional to the area of the rectangle formed by the coordinates of the point, the B-axis and the H-axis.

If $(H_i B_i)$ is plotted against B_i, its maximum value can be found and the operating point for minimum value of magnetic material obtained.

This is done in the next frame for Alcomax, one of the best known materials which contains iron, cobalt, nickel, aluminium and copper.

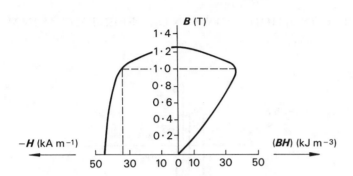

Example

An Alcomax magnet is to be used to maintain a flux density of 0.6 T across an air gap 3 mm long and 2 cm × 3 cm in cross-section. Using the data available from the above diagram, calculate the minimum volume of material required.

Solution

From the diagram we see that the flux density for maximum (BH) is 1 T. The flux in the air gap is $B_a A_a = 0.6 \times 0.02 \times 0.03 = 0.36$ mWb. Assuming no leakage or fringing, the flux in the magnet will be the same so that $B_i A_i = 0.36$ mWb and $A_i = 0.000\,36/1 = 0.000\,36$ m^2 \hfill (i)

The magnetic field strength for the air gap, $H_a = B_a/\mu_0 = 0.6/\mu_0$. The m.m.f. required for the air gap $= H_a l_a = \dfrac{0.6 \times 0.003}{4\pi \times 10^{-7}} = 1432$ A. The value of H_i for maximum (BH) is -35 kA m^{-1} (from the diagram). As we have seen, the magnitude of the m.m.f. for the air gap is the same as that for the magnet, i.e. $H_a l_a = H_i l_i$.

The length of magnet required is thus

$$\frac{H_a l_a}{H_i} = \frac{1432}{35\,000} = 0.041 \text{ m} \qquad \text{(ii)}$$

From (i) and (ii) we see that the minimum volume of material is 14.7 cm^3.

Note that the flux density required in the magnet is greater than that specified in the air gap. This is accommodated by using shaped pole pieces made of soft iron to make the cross-sectional area of the air gap larger than that of the magnet.

Other widely used permanent magnet materials are chrome steel, tungsten steel, cobalt steel and Alnico (an alloy of iron, nickel, cobalt and aluminium). These materials have values of $(BH)_{max}$ (kJ m^{-3}) of 1.6, 2.6, 7.8 and 13.6 respectively, compared with a value of 35 kJ m^{-3} for the Alcomax alloy used in the above problem.

59

BOUNDARY CONDITIONS BETWEEN DIFFERENT MEDIA IN MAGNETIC FIELDS

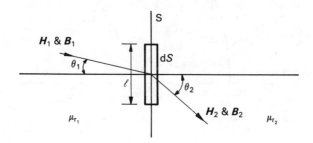

The diagram shows a cross-sectional view of a plane surface, S, separating two media having different relative permeabilities. A small part of each medium is enclosed by the thin Gaussian surface dS. The directions of the B and H vectors on either side of the boundary are as indicated on the diagram. The Gaussian surface encloses no source of flux so that the net outward flux from it is zero.

What can you deduce about the relationship between the flux densities B_1 and B_2 on either side of the boundary?

60

Their normal components are the same

The net outward flux is zero so that $B_2 \cos \theta_2 \, dS - B_1 \cos \theta_1 \, dS = 0$
i.e. $$B_1 \cos \theta_1 = B_2 \cos \theta_2$$
$B_1 \cos \theta_1$ is the normal component of B_1 (i.e. the component at right angles to the surface S). Also $B_2 \cos \theta_2$ is the normal component of B_2.

Since the Gaussian surface dS encloses no current, the path L links no current and applying Ampere's law to the path L gives

$$\oint_L H \, dl = 0$$

What can you deduce about the relationship between H_1 and H_2 on either side of the boundary? (Remember that the two short sides of the path ≈ 0.)

61

Their tangential components (i.e. those parallel to the surface) are the same

This is because $H_1 \sin \theta_1 \, l - H_2 \sin \theta_2 \, l = 0$.
The product Hl along the other two sides of the path $= 0$.
  Dividing by l gives $H_1 \sin \theta_1 = H_2 \sin \theta_2$.
$H_1 \sin \theta_1$ is the tangential component of H_1 and $H_2 \sin \theta_2$ is the tangential component of H_2.

62

THE LAW OF MAGNETIC FLUX REFRACTION

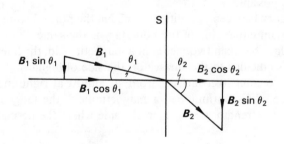

$$\tan \theta_1 = \frac{B_1 \sin \theta_1}{B_1 \cos \theta_1} = \frac{\mu_0 \mu_{r1} H_1 \sin \theta_1}{B_1 \cos \theta_1} \qquad (\text{since } B_1 = \mu_0 \mu_{r1} H_1) \qquad (i)$$

$$\tan \theta_2 = \frac{B_2 \sin \theta_2}{B_2 \cos \theta_2} = \frac{\mu_0 \mu_{r2} H_2 \sin \theta_2}{B_1 \cos \theta_1} \qquad (\text{since } B_1 \cos \theta_1 = B_2 \cos \theta_2)$$

$$= \frac{\mu_0 \mu_{r2} H_1 \sin \theta_1}{B_1 \cos \theta_1} \qquad (\text{since } H_1 \sin \theta_1 = H_2 \sin \theta_2) \qquad (ii)$$

Dividing (i) by (ii) yields $tan\, \theta_1 / \tan \theta_2 = \mu_{r1} / \mu_{r2}$.

  This is the law of magnetic flux refraction and it implies that the magnetic lines of force are more nearly perpendicular to the surface of discontinuity on the side of lower permeability.

63

THE BOUNDARY BETWEEN AIR AND A FERROMAGNETIC MATERIAL

Assume the relative permeability of air to be 1 and let this be medium 1. For an iron specimen with a relative permeability of 1000 the law of magnetic flux refraction gives

$$\frac{\tan \theta_1}{\tan \theta_2} = \frac{\mu_{r1}}{\mu_{r2}} = \frac{1}{1000} \Rightarrow \theta_1 \to 0°.$$

This means that lines of force enter ferromagnetic surfaces virtually at right angles to the surface. The higher the relative permeability of the material, the closer does θ_1 approach $0°$.

Which of the following statements are true?

(a) Permanent magnet materials have a wide hysteresis loop.
(b) The operating point for permanent magnets is in the third quadrant of the hysteresis loop where the magnetic field strength is negative.
(c) The B and H fields are in opposite directions in the air gaps of permanent magnets.
(d) In permanent magnet applications $H_a \gg H_i$.
(e) A permanent magnet keeper is made from a hard magnetic material.
(f) To design a permanent magnet for minimum volume, the (BH) product must be as small as possible.
(g) At a boundary between two different media, the normal component of magnetic flux density on either side of the boundary is the same.
(h) At a boundary between two media in a magnetic field, the lines of force are more nearly perpendicular to the surface on the side of higher μ.
(i) Lines of force enter iron surfaces from air almost at right angles.
(j) At a surface of discontinuity in a magnetic field, the tangential component of magnetic field strength is bigger on the side where the permeability is lower.

64

True: a, d, g, i

(b) They operate in the second quadrant.   (Frame 51)
(c) They are in the same direction in the air gap.   (Frame 52)
(e) Keepers are of soft magnetic material.   (Frame 55)
(f) The (BH) product must be as large as possible.   (Frame 57)
(h) The angle to the normal is smaller where μ is lower. (Frame 62)
(j) The tangential component of H on either side is the same. (Frame 61)

SUMMARY OF FRAMES 51–64

Concerning permanent magnets, which are made from hard magnetic materials:

- the **H** field and the **B** field are in opposite directions in the ferromagnetic material if the circuit includes an air gap
- the slope of the load line is $-\mu_0(l_i/l_a)$
- for minimum volume $(B_i H_i)$ must be a maximum.

At a boundary between two different media in a magnetic field:

- the normal component of **B** on either side is the same
- the tangential component of **H** on either side is the same

- $\dfrac{\tan\theta_1}{\tan\theta_2} = \dfrac{\mu_{r1}}{\mu_{r2}}$ (the law of magnetic flux refraction).

66

EXERCISES

(i)   A permanent magnet is 8 cm long and is made from a material having the following demagnetisation curve:

B (T):	0	0.4	0.7	0.82	0.9	0.94
H (kA m^{-1}):	-100	-80	-60	-40	-20	0

Determine the flux density in an air gap 5 mm long. Neglect leakage and fringing.

(ii) Determine the minimum volume of a cobalt steel permanent magnet required to maintain a flux density of 0.45 T across an air gap 2 mm in length and having a cross-sectional area of 6 cm^2. Neglect leakage and fringing. Cobalt steel has the following magnetic properties:

Value of flux density for $(BH)_{max} = 0.6$ T
Value of field strength for $(BH)_{max} = 13$ kA m^{-1}

(iii) A plane boundary separates two media, one having a relative permeability of 500 and other having a relative permeability of 6000. The **B**-vector makes an angle of 80° to the normal to the boundary on the side of lower permeability. Determine the direction of the **B**-vector in the other medium.

67

SHORT EXERCISES ON PROGRAMME 6

The numbers in brackets at the end of each question refer to the frame where the answer is to be found. All symbols have the meaning defined in the text.

1. Distinguish between paramagnetic and diamagnetic materials. (2)
2. Give an example of a ferromagnetic material. (3)
3. Define relative permeability. (6)
4. What is the unit of absolute permeability? (7)
5. What is a hysteresis loop? (13)
6. What is meant by a 'soft' magnetic material? (16)
7. What is the cause of hysteresis loss? (17)
8. Define a magnetic circuit. (23)
9. What is the unit of magneto-motive force? (23)
10. What is a major hysteresis loop? (14)
11. Give the unit of reluctance. (25)
12. What is the relationship between reluctance and permeance? (29)
13. How does the reluctance of a material vary as its length increases? (26)
14. What is meant by the mean flux path of a magnetic circuit? (33)
15. State the relationship between m.m.f., flux and reluctance. (27)
16. State the relationship between Φ, S, H and l. (25)
17. Define incremental permeability. (15)
18. Explain what is meant by magnetic leakage. (45)
19. Why can the leakage factor never be greater than unity? (45)
20. Explain the meaning of magnetic fringing. (46)
21. State the requirements for a good permanent magnet material. (16)
22. How does a keeper help a permanent magnet to retain its magnetism? (55)
23. Give the condition for minimum volume for permanent magnets. (57)
24. State the relationship between the flux density vectors on either side of a surface of discontinuity in a magnetic field. (60)
25. State mathematically the law of magnetic flux refraction. (62)

68

ANSWERS TO EXERCISES

Frame 22

(i) 7.5 W kg^{-1}; (ii) 125 W, -170 A m^{-1}, 0.54 T, 7958, 1989.

Frame 50

(i) 265×10^3 H^{-1}; (ii) 2239 A; (iii) 1.4 A.

Frame 66

(i) 0.81 T; (ii) 24.7 cm^3; (iii) 89.16° to the normal.

Programme 7

ELECTROMAGNETIC INDUCTION

1

INTRODUCTION

So far we have considered fields to be stationary and current to be steady, and we have been able to study electric fields and magnetic fields quite separately although we have noticed remarkable analogies between them. In this programme we investigate the consequences of moving fields and varying currents, and we find that there is in fact an intimate relationship between electric fields and magnetic fields.

We begin by discussing the electromagnetic induction of electromotive force upon which depends the operation of the transformer and the electric generator and without which the modern electricity supply industry would not exist.

In the course of the programme we shall learn about:

- transformer and motional electromotive forces (e.m.f.s)
- self-induced and mutually induced e.m.f.s
- Lenz's law and Fleming's right-hand rule for the direction of induced e.m.f.s
- forces on current-carrying conductors situated in uniform magnetic fields
- Fleming's left-hand rule for the direction of such forces
- images in infinitely permeable iron
- self-inductance and mutual inductance
- energy stored in magnetic fields
- forces between magnetised iron surfaces and between current-carrying coils
- torques on current-carrying coils.

When you have studied this programme you should be able to:

- calculate the magnitude and determine the direction of the e.m.f. induced in a coil or circuit as a result of the current changing in the coil itself or as a result of the current changing in some other coil or circuit
- calculate the e.m.f. induced in a conductor as a result of its movement through a magnetic field
- calculate the mutual forces between current-carrying conductors and between current-carrying conductors and iron surfaces
- calculate the coefficient of self-inductance of a coil given its physical parameters or by determining the flux linking it
- calculate the effective inductance of a number of coils connected in various configurations
- calculate the energy stored in a magnetic field in terms of the field vectors H and/or B or in terms of inductance and current
- calculate the force between magnetised iron surfaces and between current-carrying coils
- calculate the torque on a coil when it is twisted about some axis.

2

ELECTROMAGNETIC INDUCTION

In 1820 Oersted discovered that current-carrying conductors produced magnetic fields in their vicinity, and in a series of experiments in 1831 Faraday proved the converse (that an electric current could be produced by a magnetic field). He found that if two coils are fixed in position relative to each other and the current in one of them changes with time, then an electromotive force (e.m.f.) can be detected in the other.

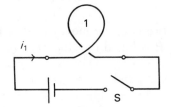 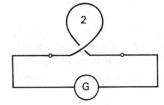

At the instant when the switch S is closed in the circuit of coil 1, the galvanometer in the circuit of coil 2 deflects momentarily. At the instant when the switch S is opened, the galvanometer deflects in the opposite direction. An e.m.f. is said to be 'induced' in coil 2 and it is found that this e.m.f. (call it E_2) is directly proportional to the rate at which the current i_1 is changing with time.

State mathematically the relationship between E_2 and i_1.

3

$$E_2 \propto \frac{di_1}{dt}$$

The current flowing in circuit 1 will set up a magnetic field whose flux is directly proportional to i_1 assuming a non-ferromagnetic medium. Some of this flux will link with coil 2. Let us call this flux Φ_{21}. This is read as 'that part of the flux produced in coil 1 which links with coil 2'.

Φ_{21} depends on the relative position of the two coils and their geometry but it will be proportional to i_1 if the medium of the field is linear.

What, then, is the relationship between E_2 and Φ_{21}?

4

$$\boxed{E_2 \propto d\Phi_{21}/dt}$$

This is obtained as follows:

Φ_{21} is directly proportional to i_1 so that the rate of change of Φ_{21} is directly proportional to the rate of change of i_1

i.e. $\dfrac{d\Phi_{21}}{dt} \propto \dfrac{di_1}{dt}$. But $E_2 \propto \dfrac{di_1}{dt}$, therefore $E_2 \propto \dfrac{d\Phi_{21}}{dt}$

If coil 1 carries a steady current I_1 and the two coils are moved relative to each other, then an e.m.f. can be detected in coil 2. It is found that the e.m.f. (call it E_2 again) is directly proportional to the relative velocity (v say) of the two coils.

State mathematically the relationship between E_2 and v.

5

$$\boxed{E_2 \propto v}$$

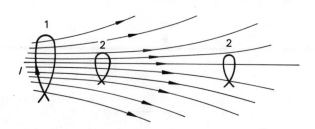

The current in coil 1 sets up a flux as before and some of it will link coil 2. We will call this Φ_{21} as before.

As the coils are moved relative to each other, so the magnitude of Φ_{21} will change and the rate of change of Φ_{21} will obviously be proportional to the relative velocity, v, of the two coils. That is

$$\frac{d\Phi_{21}}{dt} \propto v.$$

What, then, is the relationship between E_2 and Φ_{21}?

$$\boxed{E_2 \propto d\Phi_{21}/dt}$$

Comparing this answer with that in the box at the top of Frame 4, the conclusion to be drawn from the results of the two experiments is that the magnitude of the e.m.f. induced in a coil is directly proportional to the rate of change of flux linking it. This is Faraday's Law.

The e.m.f. induced in a coil where the circuits are stationary and the current is changing is often called a 'transformer' e.m.f. while the e.m.f. induced as a result of relative motion between two coils is termed a 'motional' or 'rotational' e.m.f. There is no fundamental difference between them and they are both produced by a rate of change of flux linking a conductor. Of course it is possible for both 'types' of e.m.f. to be induced in a coil at the same time. For example, if a coil moves in a field produced by a second coil carrying a changing current, the first coil will have both a transformer e.m.f. and a motional e.m.f. induced in it.

What would be the effect on the induced e.m.f. in coil 2 (Frames 2 and 5) if (i) coil 1 has N_1 turns rather than just 1, (ii) coil 2 has N_2 turns?

(i) It would be N_1 times bigger because the flux produced by coil 1 and hence Φ_{21} would be N_1 times bigger, (ii) it would be N_2 times bigger since each turn would have an e.m.f. induced in it and these add.

The e.m.f. induced in coil 2 is then given by

$$E_2 \propto \frac{d(N_2\Phi_{21})}{dt}.$$

If the flux Φ_{21} links all of the turns N_2, then N_2 is 'constant' so we may write

$$E_2 \propto N_2 \frac{d\Phi_{21}}{dt}.$$

The SI units are chosen so as to make the constant of proportionality equal to unity so that we have

$$E_2 = N_2 \frac{d\Phi_{21}}{dt}.$$

$N_2\Phi_{21}$ is called the flux linkage (λ) of circuit 2 and $d(N_2\Phi_{21})/dt$ is the rate of change of flux linkage.

Apart from the volt, what is another unit of induced e.m.f.?

8

> The weber per second $(Wb\ s^{-1})$

We have an expression for the magnitude of the induced e.m.f. We now need to consider its direction. The law governing this is known as Lenz's law and finds its origin in the law of conservation of energy. It states that the direction of the induced e.m.f. in a coil or circuit is such that it produces a current which tends to maintain constant the flux linking the coil or circuit. In other words the current, by its electromagnetic action, tends to oppose the changes which give rise to the e.m.f.

To take account of this law it is usual to write $E = -N(d\Phi/dt)$ where E is the e.m.f. induced in a coil of N turns linking a flux of Φ weber.

Is the alternative to Lenz's law thinkable?

9

> Not really, because it would mean that the current would produce a flux which would ADD to the original flux so producing a bigger e.m.f. and a bigger current and even more flux *ad infinitum.*

Example

A coil having 500 turns of wire has a direct current through it which produces a magnetic flux of 2.6 mWb.

Calculate the e.m.f. induced in the coil when the current is reversed in 200 ms.

Solution

The flux linking the coil initially = 2.6 mWb.
The flux linking it after the reversal of the current = -2.6 mWb.
The change in flux is thus 5.2 mWb and this takes place in 200 ms.

The e.m.f. induced in the coil is $-N\dfrac{d\Phi}{dt} = -500\dfrac{5.2 \times 10^{-3}}{200 \times 10^{-3}} = \underline{-13\ \text{V}.}$

A second coil, having 130 turns, is placed such that 50% of the flux produced in the first coil links with it. Calculate the e.m.f. induced in it.

$$\boxed{1.69 \text{ V}}$$

To obtain this answer, note that the flux linking the second coil is half of that linking the first, so that the flux changes by 2.6 mWb (from 1.3 mWb to −1.3 mWb) in 200 ms. The e.m.f. is then obtained by multiplying by the number of turns (130).

As a further example of the calculation of the rate of change of flux linkage, we shall consider the single turn coil shown below which has sides a and b and which is moving to the right with a velocity v in a magnetic field directed into the plane of the paper at all points.

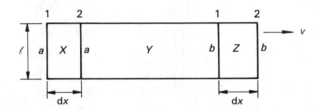

When the coil is in position 1 the flux linking it (say Φ_1) will be the flux across area X plus the flux across area Y. Thus $\Phi_1 = \Phi_X + \Phi_Y$.

When the coil has moved to position 2 the flux linking it (Φ_2) will be the flux across area Y plus the flux across area Z. Thus $\Phi_2 = \Phi_Y + \Phi_Z$.

The change in the flux linking the coil as it moves from position 1 to position 2 is therefore . . .

11

$$\boxed{(\Phi_X - \Phi_Z) \text{ (Wb)}}$$

If this change takes place in dt second then the average value of the e.m.f. induced in the coil is given by

$$E = -\frac{\mathrm{d}\Phi}{\mathrm{d}t} = -\frac{(\Phi_X - \Phi_Z)}{\mathrm{d}t} \text{ (V)} \tag{i}$$

What would be the effect if the flux in the region of side a were of the opposite sense to that in the region of side b?

12

> The sign inside the brackets of equation (i) becomes positive. This seems to be more useful because a negative sign makes the total e.m.f. zero if the fluxes Φ_X and Φ_Z are of the same magnitude.

An example of this more useful arrangement is shown below in the linearised diagram of a simple electric generator.

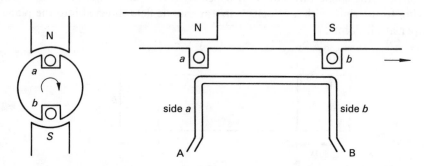

The e.m.f. induced in the coil would be zero if both poles were North or if both poles were South. Making one North and the other South ensures that there will be an e.m.f. available between the two ends A and B of the coil.

What would be the effect of having a coil of N turns rather than just 1.

13

> The e.m.f. induced or generated would be N times bigger

THE FLUX CUTTING RULE

Referring to the diagram in Frame 10, it can be seen that the flux Φ_X is the flux across the area swept out by the side a of the coil as it moves to the right a distance dx. The side a is said to have 'cut' the flux Φ_X.

Similarly, side b cuts the flux Φ_Z as it moves to the right a distance dx. The area swept out is $dx \times l$ where l is the length of each coil side and $dx = v\,dt$.

If the field in the regions of coil sides a and b is at right angles to the velocity and is of uniform flux density $\boldsymbol{B}$ there, obtain an expression for the e.m.f. induced in the coil in terms of $\boldsymbol{B}$, l and v.

14

$$E = -(\boldsymbol{B}lv \pm \boldsymbol{B}lv)$$

This is obtained as follows: Flux = flux density × area over which it acts so that $\Phi_x = \Phi_z = \boldsymbol{B}\,\mathrm{d}xl$ where $\boldsymbol{B}$ is the component of flux density which is directed at right angles to the velocity.

The e.m.f. is thus $E = -[(\boldsymbol{B}\,\mathrm{d}xl)\,\mathrm{d}t \pm (\boldsymbol{B}\,\mathrm{d}xl)\,\mathrm{d}t]$ and $\mathrm{d}x/\mathrm{d}t = v$.
The sign inside the bracket is negative if $\boldsymbol{B}$ is of the same polarity in the region of a as it is in the region of b but positive otherwise.

For a single conductor, $\underline{E = \boldsymbol{B}lv}$ where the symbols are as defined above. E, v and $\boldsymbol{B}$ are mutually at right angles and their relationship is given by Fleming's right-hand rule which says:

'If the forefinger, thumb and second finger of the right hand are held mutually at right angles with the forefinger pointing in the direction of the field and the thumb pointing in the direction of the velocity, then the second finger will be pointing in the direction of the induced e.m.f.'.

Use this rule to determine the direction of the e.m.f. induced in the conductor, C, shown in cross-section in the diagram below.

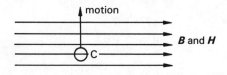

15

Into the plane of the paper

As an example of the flux cutting rule, let's find the e.m.f. generated between the wing tips of a Jumbo jet, having a wing span of 60 m, flying at 800 km h^{-1} at a height where the vertical component of the earth's magnetic field is 40 μT.

$$E = \boldsymbol{B}lv = 40 \times 10^{-6} \times 60 \times \frac{800 \times 10^3}{3600} = \underline{0.533 \text{ V}}$$

Does the answer depend upon whether the wings are of a ferromagnetic or a non-ferromagnetic material?

16

No

It might be thought that a ferromagnetic material would become magnetised thus producing more flux. This is of course true but the extra flux moves along with the wings and there is no *relative* motion.

Example

The flux linked by a coil of 100 turns varies during a complete cycle of period $T = 0.02$ s in the following manner:

From $t = 0$ to $t = T/2$, $\Phi = 0.02\,(1 - 4t/T)$ Wb.
From $t = T/2$ to $t = T$, $\Phi = 0.02\,(4t/T - 3)$ Wb.

Sketch the waveforms of flux and induced e.m.f. and determine the maximum value of the induced e.m.f.

Solution

For the flux waveform we note that the expressions are linear and draw up the following table

t (second): 0	$T/4$	$T/2$	$3T/4$	T
Φ (weber): 0.02	0	-0.02	0	0.02

For the e.m.f. waveform we use $E = -N\dfrac{\mathrm{d}\Phi}{\mathrm{d}t}$ (V).

From $t = 0$ to $t = T/2$, $\mathrm{d}\Phi/\mathrm{d}t = -0.02 \times 4/T = -4$ Wb s^{-1}
$$\therefore \quad E = -100(-4) = +400 \text{ V}$$

From $t = T/2$ to T, $\mathrm{d}\Phi/\mathrm{d}t = 0.02 \times 4/T = +4$ Wb s^{-1}
$$\therefore \quad E = -100(+4) = -400 \text{ V}$$

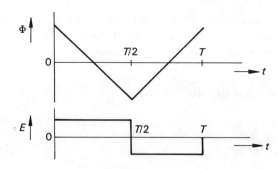

17

Which of the following statements are true?

(a) Faraday discovered that a magnetic field could not be produced by a direct current except at the instant of switching it on or off.
(b) For an e.m.f. to be induced in a conductor there must be relative motion between the conductor and magnetic flux.
(c) 'Transformer' and 'motional' e.m.f.s are fundamentally different.
(d) Faraday's law of electromagnetic induction states that the e.m.f. induced in a conductor is proportional to the rate of change of flux linkage with that conductor.
(e) If two similar coils are placed such that some of the flux produced by one of them links with the other, an e.m.f. will be induced in both of the coils if the current in one of them changes with time.
(f) Considering statement (e), the larger e.m.f. will always appear in the coil in which the current is changing.
(g) A conductor may have a transformer e.m.f. or a motional e.m.f. induced in it but not both at the same time.
(h) The weber-ampere is a unit of induced e.m.f.
(i) A straight conductor of length l carries a steady current I and moves with a velocity v in a uniform field of density B. The conductor, field and velocity are mutually at right angles. The e.m.f. induced in the conductor is given by Bll volt.
(j) For an e.m.f. to be induced in a conductor by virtue of its motion in a magnetic field, the flux density *must* be at right angles to the velocity.

18

> True: b, d, e, f

(a) There is always a magnetic field associated with a direct current.
(c) They are both caused by a rate of change in flux. (Frame 6)
(g) Both 'types' of e.m.f. can occur simultaneously. (Frame 6)
(h) The unit is the weber per second. (Frame 8)
(i) The e.m.f. induced is given by Blv. (Frame 14)
(j) There must be a *component* of B at right angles to v. (Frame 14)

19

SUMMARY (FRAMES 1–18)

Faraday's law of electromagnetic induction: the magnitude of the e.m.f. induced in a conductor is proportional to the rate of change of flux linkage with the conductor.

In magnitude, $E = \mathrm{d}(\Phi N)/\mathrm{d}t$ (transformer e.m.f.); $E = \boldsymbol{B}lv$ (motional e.m.f.).

Lenz's law: the direction of the induced e.m.f. is such as to oppose the changes causing it. To take account of this, we write $E = -\mathrm{d}(\Phi N)/\mathrm{d}t$.

20

EXERCISES

(i)   A magnetic flux of 250 μWb linking a coil of 900 turns is reversed in 0.5 s. Calculate the average value of the e.m.f. induced in the coil.

(ii) A steady current in a coil produces a magnetic flux of 450 μWb. When the current is reversed in 0.1 s there is an e.m.f. of 10 V induced in the coil. How many turns are there on the coil?

(iii) At what rate must the flux in the coil of question (ii) be reversed in order to induce an e.m.f. of 5 V in the coil?

(iv) The gauge (distance between the rails) of a railway is 150 cm. If the vertical component of the earth's magnetic field is 40 μT, calculate the reading on a voltmeter connected between the rails when a train travels along the track at 120 km h^{-1}.

(v)   A motorcar having axles of length 1.54 m travels along a road running parallel and close to the railway track of the previous question. Calculate its speed when an e.m.f. of 1.83 mV is developed between the ends of its axles.

(vi) The flux linking a 200 turn coil varies over a cycle of period 0.02 s as follows:
$\Phi = 0.02 \sin(314t)$ Wb.

Sketch the waveforms of flux and of the e.m.f. induced in the coil, and comment on their shapes and relative positions.

Calculate the maximum value of the induced e.m.f.

21

FORCES ON CURRENT-CARRYING CONDUCTORS IN UNIFORM
MAGNETIC FIELDS

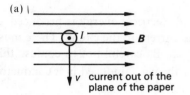 (a)

v current out of the
plane of the paper

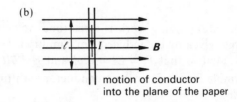 (b)

motion of conductor
into the plane of the paper

Diagrams (a) and (b) are respectively cross-sectional and plan views of a very long,
straight conductor moving with a velocity v across a magnetic field of strength B.
The 'active length' of the conductor is l.

Give an expression for the e.m.f. induced in the conductor.

22

$$E = Blv \text{ volt (see Frame 14)}$$

This e.m.f. will give rise to a current (I, say) if the conductor is part of a closed circuit.
This current will produce a magnetic field around the conductor which will combine
with the original field as shown below.

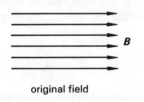 B

original field

field due to
the current in
the conductor

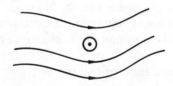

resultant, distorted field

Thinking of lines of force as being like elastic bands in that they always try to
shorten themselves, we see from the diagram on the right that the electromagnetic
force on the conductor is acting upwards. To move it downwards, therefore, requires
an external force which we shall call F. The rate at which work is done on the
conductor by this external means must equal the rate at which electrical energy is
produced in the circuit.

Remembering that electrical energy is the product of volts and amperes, write down
this energy balance equation.

23

$$\boxed{Fv = EI \text{ (W)}}$$

The rate of doing work is force times velocity, and electrical energy is produced at a rate given by the induced e.m.f. multiplied by the current in the circuit. These must be equal so that $Fv = EI = BlvI$ and $\underline{F = BlI \text{ newton.}}$ Be careful not to confuse this formula for the force on a conductor with that for the e.m.f. induced in a conductor (Blv).

A conductor of length 20 cm carries a current of 50 A and moves with a velocity of 20 m s^{-1} at right angles to a magnetic field of density 0.5 T.

Calculate (a) the e.m.f. generated in the conductor and (b) the force on it.

24

$$\boxed{\text{(a) 2 V,} \qquad \text{(b) 5 N}}$$

The direction of the electromagnetic force can be determined using Fleming's left-hand rule which says:

'If the forefinger, second finger and thumb of the left hand are held mutually at right angles with the forefinger pointing in the direction of the field and the second finger pointing in the direction of the current, then the thumb will be pointing in the direction of the electromagnetic force'.

There are then two possibilities:

(i) the conductor is moving in the direction of the electromagnetic force. This means that as a result of electrical energy being supplied, mechanical energy is being produced. The conductor is thus a part of an electric motor.

(ii) the conductor is moving against the direction in which it is being urged by the electromagnetic force. This requires an external force and means that mechanical work is being done on the conductor and this work is being converted into electrical energy. The conductor in this case is a part of an electric generator.

You must take care not to confuse Fleming's left-hand rule which is for determining the direction of the electromagnetic force on a conductor and his right-hand rule which is for determining the e.m.f. induced in it.

25

THE FORCE BETWEEN TWO PARALLEL, STRAIGHT, CURRENT-CARRYING CONDUCTORS

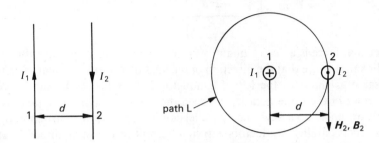

The conductors are separated by a distance which is very much greater than their radii and very much smaller than their length. They carry currents I_1 and I_2 in opposite directions. To find the force on conductor 2 we use the formula $F = BIl$ newton and let $l = 1$ m so that the force will be 'per unit length'. In this case $I = I_2$ and B is the density of the field in which conductor 2 finds itself (let this be B_2).

Now use the magnetic circuit law to determine the flux density B_2.

26

$$\boxed{B_2 = \mu_0 I_1 / 2\pi d \ \text{(T)}}$$

To obtain this result you should have applied the magnetic circuit law as follows:

$$\oint_L H_2 \, dl = I_1$$

where L is a circular path of radius d with conductor 1 as the centre and H_2 is the component of field strength in the direction of the path at every point and is constant in magnitude by symmetry.

Then

$$H_2 \oint_L dl = I_1 \Rightarrow H_2 2\pi d = I_1 \text{ and } H_2 = \frac{I_1}{2\pi d}$$

$$B_2 = \mu_0 H_2 = \frac{\mu_0 I_1}{2\pi d} \ \text{(T)}$$

So what is the force on conductor 2?

27

$$F = \frac{\mu_0 I_1 I_2}{2\pi d} \ (\text{N m}^{-1})$$

Remember that force is a vector quantity so that for complete specification we must state its direction. Note that the direction of the lines of force to the right of conductor 1 is the same as those to the left of conductor 2. It is just as though there were magnetic poles of the same polarity between the conductors and like poles repel.

We can now make up a new rule: 'parallel conductors carrying current in opposite directions repel; parallel conductors carrying current in the same direction attract'. Satisfy yourself that the second part of this rule is correct by examining the field between the conductors.

If $I_2 = I_1$ then

$$F = \frac{\mu_0 I_1^2}{2\pi d} \ (\text{N m}^{-1})$$

and putting $\mu_0 = 4\pi \times 10^{-7}$ we obtain

$$F = 2 \times 10^{-7} \frac{I_1^2}{d} \ (\text{N m}^{-1})$$

If we put $I_1 = 1$ and $d = 1$ we may define one ampere as 'that current which, when maintained in each of two very long conductors, separated by a distance of one metre, produces a mutual force of 2×10^{-7} newton per metre length between them'.

Example

Two long parallel conductors are spaced 0.04 m between centres in air and they each carry a current of 50 A in the same direction.

Calculate the mutual force between them per metre length.

Solution

Assuming that for air we may take $\mu_0 = 4\pi \times 10^{-7}$ H m^{-1} and using the formula derived above, we get

$$F = \frac{4\pi \times 10^{-7} \times 50 \times 50}{2\pi \times 0.04} \ \text{N m}^{-1} = \underline{12.5 \ \text{mN m}^{-1}}.$$

Is this a force of attraction or one of repulsion?

Attraction

IMAGES IN INFINITELY PERMEABLE IRON

We have seen (Programme 6, Frame 62) that the law of magnetic flux refraction at a boundary between two different media in a magnetic field states that

$$\frac{\mu_{r1}}{\mu_{r2}} = \frac{\tan \theta_1}{\tan \theta_2}$$

where θ_1 and θ_2 are the angles which the lines of force make with the horizontal on either side of the boundary, and μ_{r1} and μ_{r2} are the corresponding relative permeabilities.

If, therefore, $\mu_{r1} \to \infty$ and $\mu_{r2} = 1$ (air) $\tan \theta_2 = 0$. This means that magnetic flux emerges into air at right angles to the surface of infinitely permeable iron, and this leads to a method for calculating the force between a current-carrying conductor and an iron surface.

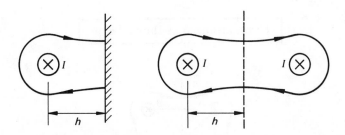

The diagram on the left shows a conductor carrying a current I and lying parallel to a sheet of infinitely permeable iron. The flux pattern is such that the lines of force enter the iron surface at right angles. Exactly the same flux pattern would be produced by the conductor and an 'image' conductor carrying the same current in the same direction as shown in the diagram on the right. The force between the conductor and the iron is therefore the same as the force between the conductor and its image. We have seen how to do this in Frame 27.

You should now be able to write down an expression for the force between the conductor and the iron in the above diagram.

29

$$F = \frac{\mu_0 I^2}{2\pi(2h)} \; (\text{N m}^{-1})$$

This is obtained by using the formula for the force between parallel conductors where $I_1 = I_2 = I$ and the separation $d = 2h$. The force will be one of attraction because the currents in the conductor and its image are in the same direction.

Now consider a conductor (A) running parallel to two sheets of infinitely permeable iron as shown in the diagram. The lines of force must enter the iron at right angles so that the flux pattern will be as shown.

How many images are required to produce a similar field pattern?

30

Three, arranged as shown below

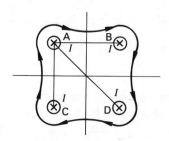

The force on the conductor in the presence of the iron surfaces may be determined by calculating the force on it in the presence of its three images. There will be a force of attraction on it due to image B acting along $\overrightarrow{AB}$; a force of attraction on it due to image C acting along $\overrightarrow{AC}$; and a force of attraction on it due to image D acting along $\overrightarrow{AD}$. These three forces must then be added vectorially to obtain the resultant.

Example

Determine the magnitude and direction of the force on the very long conductor carrying a current of 25 A in air in the two diagrams below. In each case the very long iron plates have infinite permeability.

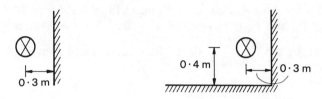

(a) The appropriate image arrangement is as shown in Frame 28 in which $I = 25$ A.

$$F = \frac{\mu_0 I^2}{2\pi d} \text{ (N m}^{-1}) = \frac{4\pi \times 10^{-7} \times 25 \times 25}{2\pi \times 0.6} \text{ N m}^{-1}$$

$$= \underline{0.208 \text{ mN m}^{-1} \text{ acting along AB}}$$

(b) The appropriate image arrangement is as shown in Frame 30 with $I = 25$ A.

The force on the conductor A due to image B acts along AB and is of magnitude

$$F_1 = \frac{4\pi \times 10^{-7} \times 25 \times 25}{2\pi \times 0.6} \text{ N m}^{-1} = 0.208 \text{ mN m}^{-1}$$

The force on the conductor due to image C acts along AC and is of magnitude

$$F_2 = \frac{4\pi \times 10^{-7} \times 25 \times 25}{2\pi \times 0.8} \text{ N m}^{-1} = 0.156 \text{ mN m}^{-1}$$

The force on the conductor due to image D acts along AD and is of magnitude

$$F_3 = \frac{4\pi \times 10^{-7} \times 25 \times 25}{2\pi \times 1.0} \text{ N m}^{-1} = 0.125 \text{ mN m}^{-1}$$

The vector diagram is as shown below in which $\phi = \tan^{-1}(0.8/0.6) = 53.13°$.

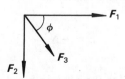

Now calculate the resultant of these three forces.

32

$$F_R = 0.382 \text{ mN m}^{-1} \text{ acting in the direction E } 42.13°\text{S}$$

The resultant horizontal force is $F_H = F_1 + F_3 \cos 53.13° = 0.283 \text{ mN m}^{-1}$.
The resultant vertical force is $F_V = F_2 + F_3 \sin 53.13° = 0.256 \text{ mN m}^{-1}$.
The resultant of F_H and F_V is $F_R = ((0.283)^2 + (0.256)^2)^{1/2} = 0.382 \text{ mN m}^{-1}$ acting
at an angle given by $\tan^{-1} \dfrac{0.256}{0.283} = 42.13°$ directed as stated above.

33

SELF-INDUCTANCE

N turns

The coil shown has a mean magnetic length l and a cross-sectional area A. The current I flowing through the N turns produces a magnetic flux Φ. If the density of the field is $\boldsymbol{B}$, the field strength is $\boldsymbol{H}$ and the medium of the field is linear, then

$$\Phi = BA = \mu_0 HA = \mu_0 \frac{NI}{l} A \left(\text{since } H = \frac{NI}{l} \right).$$

If I changes with time so, too, will Φ and

$$\frac{d\Phi}{dt} = \mu_0 \frac{NA}{l} \frac{dI}{dt}.$$

Since the flux linking the circuit is changing with time then, according to Faraday's law, an e.m.f. will be induced in it. This e.m.f. is called a 'self-induced' e.m.f. because it is induced in the coil due to the current changing in the coil itself.
As we have seen, the e.m.f. is given by

$$E = -N \frac{d\Phi}{dt} \text{ (V).}$$

Write down an expression for the e.m.f. induced in the coil in terms of the number of turns on the coil and its physical parameters.

$$E = -\frac{\mu_0 N^2 A}{l}\frac{dI}{dt}$$

This is obtained simply by substituting for

$$\frac{d\Phi}{dt} = \mu_0\frac{N A}{l}\frac{dI}{dt}$$

in the e.m.f. expression,

$$E = -N\frac{d\Phi}{dt}.$$

The constant $(\mu_0 N^2 A)/l$ is written L and is called the 'self-inductance' of the coil. The term 'self' is used because the flux is produced by the current in the coil itself.

We have, then, that

$$E = -L\frac{dI}{dt}\ (\mathrm{V}).$$

The unit of self-inductance is the Henry (H) and by putting E, L and di/dt all equal to unity in the above equation, we can define the Henry as follows: 'a coil has a self-inductance of one Henry when the current changing in it at the rate of one ampere per second induces e.m.f. of one volt'. A coil designed to have inductance is called an inductor.

For a non-ferromagnetic circuit L is a constant, but for a ferromagnetic circuit it is given by

$$\frac{\mu_0 \mu_r N^2 A}{l}\ (\mathrm{H})$$

This varies with flux density because μ_r varies with flux density.

A certain air-cored coil has a self-inductance of 2 mH. How would the inductance change if the same coil were wound on (a) a wooden former, (b) an iron former?

(a) There would be no change since μ for wood $\approx \mu_0$.
(b) The inductance would be of the order of 10^3 times bigger.

A coil of 500 turns is wound on a wooden ring having a mean diameter of 8 cm and a cross-sectional area of 1.5 cm^2. Calculate the self-inductance of the coil and the e.m.f. induced in it when a direct current of 5 A flowing in it is reversed in 0.2 s.

36

$$\boxed{187 \ \mu\text{H}; \ 9.4 \ \text{mV}}$$

The first answer is obtained by using the formula for inductance, assuming that for wood $\mu_r = 1$ and putting $N = 500$, $A = 1.5 \times 10^{-4} \ \text{m}^2$ and $l = \pi \times 0.08$ m. The second answer gives the magnitude of the induced e.m.f. and is obtained from

$$E = L \frac{\text{d}I}{\text{d}t} \qquad \text{where} \qquad \frac{\text{d}I}{\text{d}t} = \frac{10}{0.2} \ \text{A s}^{-1}.$$

We now have two expressions for the e.m.f. induced in a coil. One is in terms of the rate of change of current and the other is in terms of the rate of change of flux. Using these two expressions, find the relationship between L, N, Φ and I, making L the subject of the equation.

37

$$\boxed{L = \frac{N\Phi}{I} \ (\text{H})}$$

Since

$$E = -N \frac{\text{d}\Phi}{\text{d}t} \qquad \text{(Frame 33)}$$

and also

$$E = -L \frac{\text{d}I}{\text{d}t} \qquad \text{(Frame 34)}$$

then

$$-N \frac{\text{d}\Phi}{\text{d}t} = -L \frac{\text{d}I}{\text{d}t} \ (\text{V})$$

By integration

$$N\Phi = LI$$

so that

$$L = \frac{N\Phi}{I} \ (\text{H})$$

Inductance may therefore be measured in flux linkages per ampere. This gives us a useful method for determining inductance. We can calculate the flux linking a circuit or coil and divide by the current in it.

Calculate the inductance of a coil of 250 turns carrying a current of 8 A if this current produces a flux of 200 μWb. If the coil is air-cored and has a mean magnetic length of 10 cm, what is its cross-sectional area?

$$6.25 \text{ mH} \left(\frac{N\Phi}{I}\right); \qquad 79.6 \text{ cm}^2 \text{ (from } L = \mu N^2 A/l)$$

To illustrate the usefulness of this technique for determining inductance, we will obtain an expression for the inductance per metre of an isolated twin line (a pair of parallel conductors carrying current in opposite directions—sometimes referred to as 'go' and 'return' conductors). The arrangement is shown below in which the conductor radii are r and their separation is d ($\gg r$).

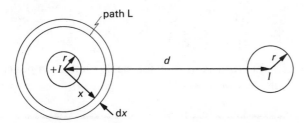

To find the inductance per metre of conductor 1 we obtain an expression for the flux linkages with it and then divided by the current, I, flowing in it. Both conductors are carrying current so that both of them will produce flux which will link conductor 1. First of all consider the flux Φ_1 linking conductor 1 due to the current $+I$ flowing in it. The steps in the calculation are:

(i)   using the magnetic circuit law, find the field strength at distance x;
(ii) determine the flux density there using $\boldsymbol{B} = \mu_0 \boldsymbol{H}$;
(iii) the flux per metre in the cylinder of thickness dx will be the flux density $\boldsymbol{B}$ times the area over which it acts (d$x \times 1$);
(iv) finally, the total flux linking conductor 1 due to its own current is then found by integrating this expression for flux with respect to the variable x from $x = r$ to $x = z$ (the large distance to the point where the field is zero).

Step (i):

$$\oint_L H_x \, \mathrm{d}l = \sum I \Rightarrow H_x \oint_L \mathrm{d}l = I$$

since $\boldsymbol{H_x}$ is constant in magnitude around path L by symmetry and the current linked is I. Thus $\boldsymbol{H_x} 2\pi x = I$ and $\boldsymbol{H_x} = I/2\pi x$ (A m^{-1}).

Step (ii): $\boldsymbol{B_x} = \mu_0 \boldsymbol{H_x} = \mu_0 I/2\pi x$ (T).

Step (iii):

$$\mathrm{d}\Phi = \frac{\mu_0 I}{2\pi x} (\mathrm{d}x \times 1) \text{ (Wb)}.$$

Now you complete step (iv).

39

$$\Phi_1 = \frac{\mu_0 I}{2\pi} \ln \frac{z}{r} \text{ (Wb)}$$

The final step was to perform the integration

$$\int_r^z \frac{\mu_0 I}{2\pi x} \, dx = \frac{\mu_0 I}{2\pi} \int_r^z \frac{dx}{x}.$$

Thus

$$\Phi_1 = \frac{\mu_0 I}{2\pi} [\ln x]_r^z = \frac{\mu_0 I}{2\pi} (\ln z - \ln r) = \frac{\mu_0 I}{2\pi} \ln \frac{z}{r}.$$

Now see if you can obtain an expression for the flux linking conductor 1 as a result of the current $(-I)$ flowing in conductor 2.

40

$$\Phi_2 = -\frac{\mu_0 I}{2\pi} \ln \frac{z}{d} \text{ (Wb)}$$

In this case the flux produced by the current $-I$ in conductor 2 is found using similar steps as before but remembering that the flux begins to link conductor 1 at a distance d from conductor 2 and stops linking it at a distance z (where the field is zero). The minus sign appears because the current is in the opposite direction to that in conductor 1.

The total flux linking conductor 1 is therefore

$$\Phi_1 + \Phi_2 = \frac{\mu_0 I}{2\pi} \ln \frac{z}{r} - \frac{\mu_0 I}{2\pi} \ln \frac{z}{d} = \frac{\mu_0 I}{2\pi} \left[\ln \frac{z}{r} - \ln \frac{z}{d} \right] \text{ (Wb)}$$

$$= \frac{\mu_0 I}{2\pi} \ln \left[\frac{z \times d}{r \times z} \right] = \frac{\mu_0 I}{2\pi} \ln \frac{d}{r} \text{ (Wb)}$$

Now inductance is equal to flux linkage per ampere so we must divide this expression by I to obtain the inductance per metre of conductor 1:

$$L = \frac{\mu_0}{2\pi} \ln \frac{d}{r} \text{ (H)}$$

Is this the complete answer? Or have we overlooked some flux linkages?

41

There will be some internal flux linkage, that
is to say within the material of the conductor.

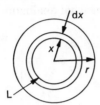

To obtain an expression for this 'internal' flux linkage, consider an annulus of the
conductor of radius x ($<r$) and radial thickness dx. Assume that the current is
uniformly distributed over the cross-section of the conductor so that at radius x the
path L links a fraction $(x/r)^2$ with I.

Apply the magnetic circuit law to path L,

$$H_x \oint_L dl = (x/r)^2 I.$$

$$\therefore \quad H_x \times 2\pi x = (x/r)^2 I \quad \text{and} \quad H_x = Ix/2\pi r^2 \ (\text{A m}^{-1})$$

The flux in the annulus per metre length is given by $d\Phi = B_x \times (dx \times 1)$ i.e.
$d\Phi = \mu H_x \, dx = \mu(Ix/2\pi r^2) \, dx$ and this flux links $(x/r)^2$ of 1 turn.

Thus flux linkages with the annulus,

$$d(\Phi N) = \frac{\mu Ix}{2\pi r^2} \times \frac{x^2}{r^2} \, dx = \frac{\mu Ix^3}{2\pi r^4} \, dx$$

The total internal flux linkages are therefore

$$\int_0^r \frac{\mu Ix^3}{2\pi r^4} \, dx \ (\text{Wb})$$

$$= \frac{\mu I}{2\pi r^4} \int_0^r x^3 \, dx = \frac{\mu I}{2\pi r^4} \frac{r^4}{4} = \frac{\mu I}{8\pi} \ (\text{Wb})$$

The complete expression for the flux linkage with conductor 1 is therefore

$$\Phi_{\text{int}} + \Phi_{\text{ext}} = \frac{\mu I}{8\pi} + \frac{\mu_0 I}{2\pi} \ln \frac{d}{r} \ (\text{Wb})$$

The inductance per metre length is thus

$$\frac{\mu}{8\pi} + \frac{\mu_0}{2\pi} \ln \frac{d}{r} \ (\text{H}).$$

What is μ in this expression?

42

The permeability of the conductor material

The most important effect of inductance in a circuit is that it prevents current from changing instantaneously. It is somewhat analogous to inertia in mechanical systems.

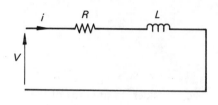 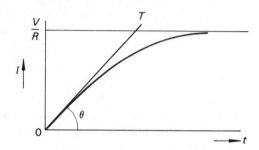

If a d.c. voltage is suddenly applied to a coil having inductance L and resistance R, Kirchhoff's second (voltage) law tells us that at a time t second the voltage is given by the differential equation

$$v = iR + L\frac{di}{dt}$$

Solving for i gives $i = I(1 - e^{-Rt/L})$ ampere where I is the final (Ohm's Law) value of the current (V/R). The graph of this is shown above.

What determines how quickly the current reaches its 'final' value?

43

The amount of inductance in the circuit

Mathematically, the current never reaches its final value and we use the idea of a 'time constant' to give an indication of the rate of growth of current.

The time constant is L/R second and is the time taken for the current to reach 63.2% of its Ohm's Law value. If the current continued to rise at the initial rate, it would reach this value in a time equal to the time constant and this is entirely dependent on the inductance.

To see that this is so, note from the equation in the previous frame that at $t = 0$, $iR = 0$ and $v = L \times$ rate of initial growth of current.

Thus $V/L = I/T$ and $\tan \theta = V/L$.

44

Example

A d.c. voltage of 100 V is suddenly applied to a coil having a resistance $R = 5\ \Omega$ and an inductance $L = 2.5$H. Calculate

(a) the initial rate of rise of current,
(b) the time constant,
(c) the 'final' value of the current.

Solution

(a) Initial rate of rise of current $= \dfrac{V}{L} = \dfrac{100}{2.5} = \underline{40\ \text{A s}^{-1}}.$

(b) The time constant $= \dfrac{L}{R} = \dfrac{2.5}{5} = \underline{0.5\ \text{s}}.$

(c) The final value of the current $= \dfrac{V}{R} = \dfrac{100}{5} = \underline{20\ \text{A}}.$

45

MUTUAL INDUCTANCE

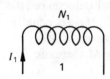

If the coil 1, having N_1 turns, has a current I_1 flowing through it, then a flux will be set up by the m.m.f. $N_1 I_1$. Call this flux Φ_1.

If a second coil, having N_2 turns, is placed such that some of the flux Φ_1 links with it, then the two coils are said to be mutually coupled.

Let the flux linking the N_2 turns of coil 2 be Φ_{21}. Because Φ_{21} is simply a part of Φ_1 we can write

$$\Phi_{21} \propto N_1 I_1$$

or $\qquad\qquad \Phi_{21} = kN_1 I_1 \text{ where } k \text{ is a constant}$

If I_1 is changing with time, Φ_1 and Φ_{21} will also be changing with time.

We have seen that there is an e.m.f. induced in coil 1 due to the rate of change of flux linkage with it and we called this a 'self-induced' e.m.f.

Write down an expression for this e.m.f.

46

$$E_1 = -N_1 \, d\Phi_1/dt \ (V)$$

There are also changing flux linkages with the coil 2 so in accordance with Faraday's law there will be an e.m.f. induced in it, too. Call this E_2. An e.m.f. induced in a coil due to the current changing in another coil is called a 'mutually induced e.m.f.'. The changing flux Φ_{21} linking coil 2 is produced by the changing current I_1 in coil 1. The e.m.f. E_2 is therefore a mutually induced e.m.f.

Write down an expression for E_2.

47

$$E_2 = -N_2 \, d\Phi_{21}/dt \ (V)$$

Since $\Phi_{21} = kN_1 I_1$ then

$$\frac{d\Phi_{21}}{dt} = kN_1 \frac{dI_1}{dt} \qquad \text{so that} \qquad E_2 = -kN_1 N_2 \frac{dI_1}{dt} \ (V).$$

$kN_1 N_2$ is written M_{21} and is called the coefficient of mutual inductance of coil 2 with respect to coil 1. It can be shown that the mutual inductance (the 'coefficient' bit is usually omitted) of coil 1 with respect to coil 2, M_{12}, is equal to M_{21}. We simply say that the mutual inductance between the two coils is M.

We have that $\Phi_{21} = kN_1 I_1$ and that $kN_1 N_2 = M$ therefore

$$kN_1 = \frac{M}{N_2} \qquad \text{and} \qquad \Phi_{21} = \frac{MI_1}{N_2}$$

It follows that

$$M = \frac{\Phi_{21} N_2}{I_1}$$

flux linkages per ampere.

The unit of mutual inductance is thus the same as that of self-inductance. If there is a current I_2 in coil 2 which is changing and producing a changing flux Φ_2 of which Φ_{12} links with coil 1, then the mutual inductance between the two coils is given by

$$M = \frac{\Phi_{12} N_1}{I_2} \ (\text{Wb A}^{-1}).$$

48

COEFFICIENT OF COUPLING

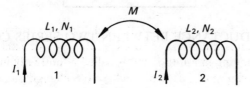

For the coil 1 we have

$$L_1 = \frac{N_1 \Phi_1}{I_1} \qquad \text{so that} \qquad \Phi_1 = \frac{L_1 I_1}{N_1} \text{ (Wb)}$$

also

$$M = \frac{N_2 \Phi_{21}}{I_1} \qquad \text{so that} \qquad \Phi_{21} = \frac{M I_1}{N_2} \text{ (Wb)}.$$

Now Φ_{21} is that part of Φ_1 which links coil 2 so we can say that

$$\Phi_1 > \Phi_{21} \quad \text{i.e.} \quad \frac{L_1 I_1}{N_1} > \frac{M I_1}{N_2} \quad \text{and} \quad \frac{L_1}{N_1} > \frac{M}{N_2}$$

By considering the flux produced by the current I_2 in coil 2 and the part of that which links coil 1, obtain the relationship between L_2, M, N_1 and N_2.

$$\boxed{L_2/N_2 > M/N_1}$$

Multiplying these two inequalities we see that

$$\frac{L_1 L_2}{N_1 N_2} > \frac{M^2}{N_1 N_2}$$

and multiplying both sides by $N_1 N_2$ gives $L_1 L_2 > M^2$.

It follows that $M < (L_1 L_2)^{1/2} = k(L_1 L_2)^{1/2}$ where k is a 'coefficient of coupling' and cannot be greater than 1. If $k \to 1$, the coils are said to be closely coupled whereas if $k \to 0$, they are said to be loosely coupled.

Calculate the mutual inductance between a coil having a self-inductance $L_1 = 20$ mH and a coil having a self-inductance $L_2 = 50$ mH which is wound on top of but insulated from it. Assume perfect coupling (i.e. $k = 1$).

50

$$\boxed{31.62 \text{ mH } (L_1 L_2)^{1/2}}$$

THE MUTUAL INDUCTANCE BETWEEN TWO SERIES CONNECTED COILS

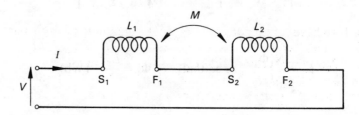

In the circuit shown, coil 1 has a self-inductance of L_1 and the current I flowing through its N_1 turns produces a flux of Φ_1 weber. The corresponding quantities for coil 2 are L_2, I, N_2 and Φ_2.

The total flux linking coil 1 is $(\Phi_1 + \Phi_{12})$ weber where Φ_{12} is part of Φ_2. This assumes that the coils are wound in such a way that the fluxes are additive. Reversing the connections to one of the coils would reverse the direction of the current through it and therefore the polarity of the flux produced in it. The fluxes in the two coils would then be subtractive. The 'starts' and 'finishes' of the coil windings are identified by 'S' and 'F'. The total flux linkage with coil 1 is given by:

$$N_1 \Phi_1 + N_1 \Phi_{12} = L_1 I + MI = (L_1 + M)I \text{ (Wb)}$$

If the current is changing with time, write down expressions for the magnitudes of the e.m.f.s. induced in coil 1.

51

$$\boxed{E_{11} = L_1 \, dI/dt \text{ (V)}; \qquad E_{12} = M \, dI/dt \text{ (V)}}$$

E_{11} is a self-induced e.m.f. and is due to the current changing in coil 1 itself. E_{12} is a mutually induced e.m.f. due to the current changing in coil 2. The total e.m.f. induced in coil 1 is therefore given by

$$E_1 = E_{11} + E_{12} = (L_1 + M)\frac{dI}{dt} \text{ (V)}$$

If Φ_{21} is that part of the flux Φ_1 which links the N_2 turns of coil 2, obtain expressions for (a) the total flux linkages with coil 2 and (b) the total e.m.f. induced in coil 2.

(a) $(N_2\Phi_2 + N_2\Phi_{21})$ weber, (b) $(L_2 + M)\,\mathrm{d}I/\mathrm{d}t$ volt

The magnitude of the total e.m.f. induced in this 'series-aiding' combination (as this arrangement is called) is therefore given by

$$E_1 + E_2 = (L_1 + L_2 + 2M)\,\mathrm{d}I/\mathrm{d}t \; (\mathrm{V})$$

A single coil which would take the same current from the same supply as the series combination would need to have a self-inductance of $(L_1 + L_2 + 2M)\,(\mathrm{H})$ and this is called the 'effective' inductance of the series combination.

If we now reverse the connections to one of the coils, say coil 1, the flux produced by the current in its N_2 turns will be reversed in sign so that it will 'oppose' the flux produced in coil 1.

The total flux linked with coil 1 becomes $\lambda_1 = (N_1\Phi_1 - N_1\Phi_{12})\,(\mathrm{Wb})$

$$\therefore \quad \lambda_1 = (L_1 I - MI) = (L_1 - M)I \; (\mathrm{Wb})$$

The magnitude of the e.m.f. induced in coil 1 is thus $(L_1 - M)\,\mathrm{d}I/\mathrm{d}t\,(\mathrm{V})$.

Similarly the magnitude of the e.m.f. induced in coil 2 is $(L_2 - M)\,\mathrm{d}I/\mathrm{d}t\,(\mathrm{V})$. The total e.m.f. induced in this 'series-opposing' combination is therefore $(L_1 + L_2 - 2M)\,\mathrm{d}I/\mathrm{d}t$ and the effective inductance is $(L_1 + L_2 - 2M)\,(\mathrm{H})$.

Two coils having inductances of 30 mH and 50 mH are connected in series and are so placed that the mutual inductance between them is 31 mH. Calculate the effective inductance of the series combination.

142 mH or 18 mH $(L_1 + L_2 \pm 2M)$

You were not able to give one answer because you weren't told how the coils were wound. In order to avoid having to use $\pm$ throughout the solutions to this type of problem, we make use of 'the dot notation'.

The rules are quite simple to apply:

(i)   plate a dot at one end of each coil (it doesn't matter which end);
(ii) indicate by means of an arrow the direction of the self-induced e.m.f. in each coil—this is always opposing the current;
(iii) if the self-induced e.m.f. in one coil points to its dotted end, then the mutually induced e.m.f. in the other coil points to *its* dotted end.

The next frame gives an example of the use of the dot notation.

54

Example

A coil having a self-inductance of 20 mH is connected in series with a second coil of self-inductance 40 mH such that there is a mutual inductance of 25 mH between them. Determine the total e.m.f. induced in the coils when a current changing at a rate of 5 A s^{-1} is passed through them.

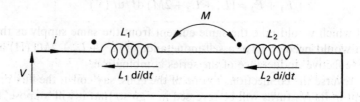

The arrangement is shown above and to use the dot notation we place a dot at the left-hand end of each coil (arbitrarily chosen).

The magnitudes of the self-induced e.m.f.s are $L_1 \, di/dt$ and $L_2 \, di/dt$ and their directions are indicated in accordance with step (ii).

The magnitude of the mutually induced e.m.f. in each coil is $M \, di/dt$.

Now decide, for each coil, the direction of the mutually induced e.m.f.

55

Towards the dotted end in both coils

This is because in each case the direction of the self-induced e.m.f. in the other coil is towards *its* dotted end.

All four induced e.m.f.s are therefore acting in the same direction so that the total e.m.f. is $(L_1 + L_2 + 2M) \, di/dt = (20 + 40 + 50) \times 5 = 550$ mV.

If we now reverse the connections to both of the coils there should be no net effect because we will have reversed the direction of the flux in both of them, so that if they were aiding before they will still be aiding. This reversal is shown by placing the dot at the other end of each of the coils. Both self-induced e.m.f.s are now directed towards the undotted end of each coil so the mutually induced e.m.f. in each coil will be directed to the undotted ends. All four e.m.f.s are thus directed in the same direction as before.

If we reverse the connections to one of the coils, say coil 2, we can show this by placing the dot at the other end of that coil.

What would be the total e.m.f. induced in this series opposing arrangement?

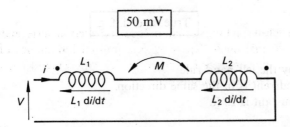

$$50 \text{ mV}$$

The arrangement is shown above and we note that the direction of the induced e.m.f. in coil 1 is towards its dotted end. The direction of the mutually induced e.m.f. in coil 2 is therefore towards its dotted end. The self-induced e.m.f. in coil 2 is directed towards its undotted end so that the mutually induced e.m.f. in coil 1 is towards its undotted end. The two mutually induced e.m.f.s are thus in the opposite direction to the two self-induced e.m.f.s. The total e.m.f. in the series combination is therefore $(L_1 + L_2 - 2M)\,di/dt$.

Could the same effect have been achieved without actually reversing the connections to either coil?

57

Yes, simply by turning over coil 2

Now decide which of the following statements are true:

(a) A straight conductor of length l carrying a current I moves with a velocity y in a uniform field of density B. B, I and v are mutually are right angles. The conductor experiences a force of BIl (N).

(b) The direction of the force on the conductor in (a) is given by Fleming's right-hand rule.

(c) Two parallel conductors carrying current in the same direction experience a mutual force of attraction.

(d) If three current-carrying coils are placed such that there is mutual magnetic coupling between all pairs, there will be a total of three self-induced e.m.f.s and six mutually induced e.m.f.s

(e) The magnetic image of a conductor in a given plane carries current in the opposite direction to the conductor.

(f) The unit of inductance is the Henry which is equal to the weber per ampere.

(g) Inductance retards the growth of current in a circuit.

(h) The time constant of a coil is given by R/L (s).

58

True: a, c, d, f, g

(b) It is given by the left-hand rule.   (Frame 24)
(e) The image current is in the same direction.   (Frame 28)
(h) The time constant is L/R.   (Frame 43)

59

SUMMARY OF FRAMES 21–58

Quantity	Symbol	Unit	Abbreviation
Self-inductance	L	Henry	H
Mutual inductance	M	Weber per ampere	Wb A^{-1}

Formulae

$F = BIl$ Force on a current-carrying conductor

$F = \dfrac{\mu I_1 I_2}{2\pi d}$ Magnitude of the mutual force between parallel current carrying conductors

$L = \dfrac{\mu N^2 A}{l}$ Inductance of a magnetic circuit

$L = \lambda/I$ flux linkages per ampere flowing in a circuit

$M = \lambda_1/I_2$ flux linkages with one circuit per ampere flowing in another circuit

$M = k\,(L_1 L_2)^{1/2}$ Mutual inductance between two coils

$k = $ the coefficient of coupling: $(0 < k < 1)$

$L_1 + L_2 \pm 2M$ The effective inductance of two coils connected in series aiding ($+$) or series opposing ($-$)

$E_2 = M\dfrac{dI_1}{dt}$ The mutually induced e.m.f. in a circuit

$E = -L\dfrac{dI}{dt}$ The self-induced e.m.f. in a circuit

$L/R = T$ The time constant of a circuit containing resistance and inductance

60

EXERCISES

(i)   Calculate the force on a conductor having an active length of 0.5 m and carrying a current of 40 A when it lies in a magnetic field of density 0.23 T.

(ii)   Two long parallel conductors each carry 60 A in the same direction in air. They are spaced 0.5 m between axes. Calculate the force between them per metre length and explain, with the aid of a diagram, whether the force is one of attraction or repulsion.

(iii)   A long conductor carries a current of 250 A and runs parallel with and 0.1 m from a sheet of infinitely permeable iron. Calculate the force per metre length on the conductor.

(iv)   A non-ferromagnetic ring has a mean diameter of 0.2 m and a cross-sectional area of 3 cm². It is uniformly wound with 350 turns of wire. Calculate the inductance of the coil.

(v)   A coil of 400 turns is wound uniformly over the top of the coil in question (iv) above. Calculate the mutual inductance between the coils, stating any assumptions made.

(vi)   Calculate the effective inductance of the two coils in the previous question when they are connected in series aiding.

(vii)   Two similar coils are placed such that there is a coupling factor of 0.2 between them. When they are connected in series aiding, the effective inductance is 65 μH. Calculate the self-inductance of each coil.

(viii) Determine the inductance per metre run of an isolated twin line having conductors of radius 1.2 cm separated by 0.8 m.

(ix)   A coil having a resistance of 2 Ω and an inductance of 0.5 H is suddenly connected to a steady voltage of 100 V. Calculate the initial rate of rise of current in the circuit and its time constant.

61

ENERGY IN THE MAGNETIC FIELD

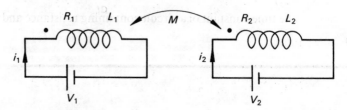

When the batteries V_1 and V_2 are connected to coils 1 and 2 respectively, the currents i_1 and i_2 will begin to grow towards their final values at rates depending on the ratio of inductance to resistance in each coil. Note that there is one self-induced e.m.f., one mutually induced e.m.f. and one resistive volt drop in each circuit. With the dots placed as shown, write down Kirchhoff's second (voltage) law for circuit 1.

62

$$V_1 - i_1 R_1 - L_1 \frac{di_1}{dt} - M \frac{di_2}{dt} = 0$$

Applying the dot notation indicates that the mutually induced e.m.f. is in the same direction as the self-induced e.m.f. Taking the clockwise direction around the circuit, V_1 is positive and all other terms are negative.

We can rewrite this as: $V_1 = i_1 R_1 + L_1 \frac{di_1}{dt} + M \frac{di_2}{dt}$ (V)

Similarly for circuit 2: $V_2 = i_2 R_2 + L_2 \frac{di_2}{dt} + M \frac{di_1}{dt}$ (V)

The battery of circuit 1 is supplying energy to the system at a rate of $V_1 i_1$ joule per second (watt) and the battery of circuit 2 is supplying energy to the system at a rate of $V_2 i_2$ joule per second.

Now $V_1 i_1 = i_1^2 R_1 + L_1 i_1 \frac{di_1}{dt} + M i_1 \frac{di_2}{dt}$ (J s^{-1})

and $V_2 i_2 = i_2^2 R_2 + L_2 i_2 \frac{di_2}{dt} + M i_2 \frac{di_1}{dt}$ (J s^{-1})

The total energy being supplied to the system is therefore at the rate

$$i_1^2 R_1 + i_2^2 R_2 + L_1 i_1 \frac{di_1}{dt} + L_2 i_2 \frac{di_2}{dt} + M i_1 \frac{di_2}{dt} + M i_2 \frac{di_1}{dt} \text{ (J s}^{-1})$$

This can be rewritten as

$$i_1^2 R_1 + i_2^2 R_2 + \frac{d}{dt} \left[\frac{1}{2} L_1 i_1^2 + \frac{1}{2} L_2 i_2^2 + M i_1 i_2 \right] \text{ (J s}^{-1})$$

because if we differentiate the inductance terms with respect to the variables i_1 and i_2 we return to the previous expression.

If both i_1 and i_2 are zero at an instant $t = 0$, give an expression for the energy supplied to the system at time T when both currents have reached their final values.

63

$$\boxed{\int_0^T (i_1^2 R_1 + i_2^2 R_2) \, dt + \int_0^T \frac{d}{dt} \left[\frac{1}{2} L_1 i_1^2 + \frac{1}{2} L_2 i_2^2 + M i_1 i_2 \right] dt \text{ (J)}}$$

The $i^2 R$ terms represent energy which has been converted into heat and has been 'lost'. The other terms, involving inductance, represent energy which has been stored in the system and which is recoverable.

Energy stored, then, is given by

$$\int_0^T \frac{d}{dt} \left[\frac{1}{2} L_1 i_1^2 + \frac{1}{2} L_2 i_2^2 + M i_1 i_2 \right] dt = \frac{1}{2} L_1 I_1^2 + \frac{1}{2} L_2 I_2^2 + M I_1 i_2 \text{ (J)}$$

This result was obtained for a system of just two circuits. For a general system of n circuits the energy stored is given by

$$\sum_{a=1}^n \frac{1}{2} L_a I_a^2 + \sum_{a=1, b=1, a \neq b}^n M_{ab} I_a I_b \text{ (J)}$$

If only one circuit carries current, then $n = 1$ and the energy stored is simply given by

$$W = \frac{1}{2} L i^2 \text{ (J)}$$

What is the electrostatic field analogy of this expression?

64

$$\text{The energy stored in a capacitor} = \frac{1}{2}CV^2 \text{ (J)}$$

In the electrostatic field case we also expressed the energy stored in terms of the field vectors E and/or D. Let's do the same for the magnetic field by expressing the energy stored in terms of the field vectors B and/or H. To do this we consider a coil of area A and mean magnetic length l carrying a current I. If the inductance of the coil is L henry, then the energy stored is given by

$$W = (1/2)LI^2 \text{ (J)}$$

$$= \frac{1}{2}\frac{N\Phi}{I}I^2 \quad \text{since} \quad L = \frac{N\Phi}{I} \text{ (Wb A}^{-1}\text{)}$$

$$= (1/2)NI\Phi = (1/2)NIBA$$

$$= \frac{1}{2}\frac{NI}{l}lBA = \frac{1}{2}HBlA \text{ (J)}$$

Since Al is the volume of the field, the energy stored may be expressed as $(1/2)HB$ joule per cubic metre.

Express this in terms of (a) H and μ, (b) B and μ.

65

$$\text{(a) } \mu H^2/2, \text{ (b) } B^2/2\mu \text{ (J m}^{-3}\text{)}$$

These expressions are also analogous to those of the electrostatic field.

Example

A steel ring has a cross-sectional area of 8×10^{-4} m^2 and a radial air gap of length 2 mm. Calculate the energy stored in the field when a coil wound on the ring produces a total flux of 720 μWb. Allow a leakage factor of 1.2.

The solution is given in the next frame. Try it yourself, first.

358 mJ

This is obtained by using the formula

$$W = \frac{1}{2}\frac{B^2}{\mu} \times \text{the volume of the field.}$$

In this case $B = \Phi/A$ where $\Phi = (720/1.2)$ μWb and $A = 8 \times 10^{-4}$ m^2. The volume of the field is the length of the air gap multiplied by its cross-sectional area $= (2 \times 10^{-3} \times 8 \times 10^{-4})$ m^3.

Now here's an example using the formula involving inductances:

A non-magnetic ring of mean diameter 0.4 m and cross-sectional area 3×10^{-4} m^2 is uniformly wound with a coil of 500 turns. A second ring, similar to the first, is uniformly wound with 250 turns and placed so that the coupling factor is 0.6. Calculate the energy stored in the system when the first coil carries a current of 5 A and the second coil carries a current of 10 A.

The energy stored in the system is given by

$$W = \frac{1}{2}L_1 I_1^2 + \frac{1}{2}L_2 I_2^2 + M I_1 I_2.$$

We therefore need to calculate L_1, L_2 and M.

$$L_1 = (\mu_0 N^2 A)/l = (4\pi \times 10^{-7} \times 500^2 \times 0.3 \times 10^{-3})/0.4\pi = \underline{75 \ \mu H}$$

Now you calculate L_2 and M.

67

$L_2 = 18.75 \ \mu H, \ M = 22.5 \ \mu H$

$L_2 = (250/500)^2 L_1$ and $M = 0.6(L_1 L_2)^{1/2}$.
The energy stored is thus $[1/2(75 \times 5^2 + 18.75 \times 10^2) + 22.5 \times 5 \times 10] \ \mu J = \underline{3000 \ \mu J}$.

The answer is given in μJ because the inductances are in μH.

The maximum electric field strength in air is 3×10^6 V m^{-1}, so that the maximum energy density is about 40 J m^{-3} ($\varepsilon_0 E^2/2$). Magnetic flux densities in air gaps of machines are commonly 1 T, so that the energy density is of the order of 400×10^3 J m^{-3}!

68

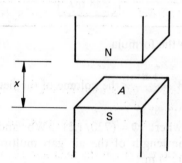

FORCES BETWEEN MAGNETISED IRON SURFACES

The diagram shows oppositely magnetised poles. The area of the pole faces is A and they are separated by a uniform distance x. If we assume the field to be uniform between the pole faces and ignore leakage and fringing effects, then the energy stored in the field is given by

$$W = \frac{B^2 A x}{2\mu_0} \text{ (J)}.$$

Since the poles are oppositely magnetised, there will be a force of attraction between them so that to separate them will require an external force, F say.

If the separation is increased by a small amount dx, the energy stored will increase because the volume of the field will have increased.

Write down an expression for the increase in stored energy.

69

$$\boxed{\frac{B^2 A \, dx}{2\mu_0} \text{ (J)}}$$

This increase in stored energy must equal the work done, $F \, dx$, by the external force in moving the poles against the magnetic force tending to pull them together

$$\text{i.e.} \quad F \, dx = \frac{B^2 A \, dx}{2\mu_0} \text{ (J)}$$

Dividing both sides by dx we get

$$F = \frac{B^2 A}{2\mu_0} \text{ (N)}.$$

Using the relationship $B = \mu_0 H$, express the force in terms of (a) B and H and (b) H and μ_0.

$$\boxed{\text{(a) } F = (1/2)BHA \text{ (N), (b) } F = (1/2)\mu_0 H^2 A \text{ (N)}}$$

Example

A pair of oppositely magnetised poles each have a face area of 10^{-3} m^2 and are separated by a uniform distance of 2 cm.

A uniform flux of 800 μWb exists between the poles and the field outside the space between the poles is negligible.

Calculate (a) the energy stored (W) in the field and (b) the force of attraction (F) between the poles.

Solution

The flux density between the poles is $800 \times 10^{-6}/10^{-3} = 800$ mT.

(a) $W = \dfrac{B^2}{2\mu_0}(Ax)\text{ (J)} = \dfrac{(800 \times 10^{-3})^2 \times 10^{-3} \times 0.02}{2 \times 4\pi \times 10^{-7}} = \underline{5.09 \text{ J.}}$

(b) $F = \dfrac{B^2 A}{2\mu_0}\text{ (N)} = \dfrac{(800 \times 10^{-3})^2 \times 10^{-3}}{2 \times 4\pi \times 10^{-7}} = \underline{254 \text{ N.}}$

71

FORCES BETWEEN CURRENT-CARRYING CIRCUITS

Current-carrying coils have the equivalent of magnetic poles at their ends so that they will exhibit mutual forces of repulsion or attraction depending on the relative directions of the currents.

The system shown consists of two coils of self-inductances L_1 and L_2 so positioned that the mutual inductance is M.

Write down an expression for the energy stored when the currents are as shown.

72

$$W = ((1/2)L_1I_1^2 + (1/2)L_2I_2^2 + MI_1I_2) \, (\text{J})$$

The dots indicate that the coils are wound in the same sense and the currents are such that the adjacent polarities are opposite. There will be a force of attraction between the coils. If an external force is used to increase the separation slightly against the magnetic force tending to pull them together, then the work done by the external force will be equal to the increase in the stored energy of the system. Assuming that the currents are maintained constant, will L_1, L_2 and M change as the coils are separated?

73

Neither L_1 nor L_2 will change but M will change

If the coils are separated by a distance dx and if this causes the mutual inductance to change by an amount dM, then the work done by the external force is $F \, dx$ joule and the increase in stored energy in the field is dMI_1I_2 joule, so that $F \, dx = dMI_1I_2$ and $F = I_1I_2 \, dM/dx$ (N).

What would be the effect of twisting one of the coils through a small angle θ about some chosen axis? (Assume that the currents are unchanged.)

74

Again, M would change but L_1 and L_2 would not

This time an external *torque* would be required and the work done by it will equal the change in stored energy. If a twist of $d\theta$ results in a change in mutual inductance amounting to dM, then the increase in stored energy is dMI_1I_2 joule and the work done by the external torque is $T \, d\theta$ joule. These must be equal so that $T \, d\theta = dMI_1I_2$ and $T = I_1I_2 \, dM/d\theta$ (N m).

Example

Two circular coils, separated by a distance x, share a common axis. Coil 1 has a radius R and carries I_1 ampere while coil 2 has a radius r and carries I_2 ampere. The coils have N_1 and N_2 concentrated turns respectively. Obtain expressions for (a) the force between the coils and (b) the torque on coil 2 when it is twisted through an angle θ.

Solution

(a)

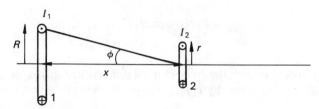

In order to apply the formula

$$F = \frac{dM}{dx} I_1 I_2 \text{ (N)}$$

derived in Frame 73 we need an expression for M, which means obtaining the flux linkages with one of the coils, say coil 2, and dividing by the current in the other. The flux linking coil 2 is the flux density there (B_2, say) times the area of the coil. The flux density B_2 is found from $B = \mu_0 H$.

Can you remember the expression for the magnetic field strength, H, on the axis of a coil such as coil 1?

$$\boxed{H = \frac{N_1 I_1}{2R} \sin^3\phi \ (\text{A m}^{-1}) \text{ (see Programme 5, Frame 44)}}$$

... and $B = \mu_0 H$.

In this case

$$\sin\phi = \frac{R}{(R^2 + x^2)^{1/2}} \qquad \text{so that} \qquad B_2 = \frac{\mu_0 N_1 I_1}{2R} \times \frac{R^3}{(R^2 + x^2)^{3/2}} \text{ (T).}$$

Obtain an expression for Φ_2, the flux linking coil 2.

77

$$\Phi_2 = \frac{\mu_0 N_1 I_1}{2R} \times \frac{R^3}{(R^2 + x^2)^{3/2}} \pi r^2 \ (\text{Wb})$$

This is simply B_2 times the area over which it acts (πr^2).

The flux linkages with coil 2 per ampere in coil 1 give the mutual inductance between the coils, i.e.

$$M = \frac{\Phi_2 N_2}{I_1} = \frac{\mu_0 N_1 N_2 \pi r^2 R^2}{2(R^2 + x^2)^{3/2}} \ (\text{H}).$$

To find the force between the coils we must differentiate this expression with respect to x and multiply the result by $I_1 I_2$.

$$\frac{\mathrm{d}M}{\mathrm{d}x} = (\mu_0 N_1 N_2 \pi r^2 R^2 / 2)(-(3/2)((R^2 + x^2)^{-5/2} \times 2x)) = -\frac{3\mu_0 N_1 N_2 \pi r^2 R^2 x}{2(R^2 + x^2)^{5/2}} \ (\text{N})$$

Finally

$$F = \frac{\mathrm{d}M}{\mathrm{d}x} I_1 I_2 = -\frac{3\mu_0 N_1 N_2 \pi r^2 R^2 x I_1 I_2}{2(R^2 + x^2)^{5/2}} \ (\text{N}).$$

(b) *The torque on coil 2*

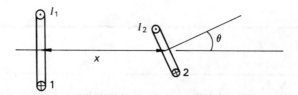

If coil 2 is twisted through an angle θ as shown, then the flux density, the flux linking it and the mutual inductance between the coils will all be $\cos \theta$ times their previous values

i.e.
$$M = \frac{\mu_0 N_1 N_2 \pi r^2 R^2}{2(R^2 + x^2)^{3/2}} \cos \theta \ (\text{H})$$

Now complete the problem and write down an expression for the torque on coil 2.

$$T = -\frac{\mu_0 N_1 N_2 \pi r^2 R^2 I_1 I_2}{2(R^2 + x^2)^{3/2}} \sin \theta \; (\text{N m})$$

This is obtained by differentiating the expression for M with respect to θ and then multiplying by $I_1 I_2$.

Which of the following statements are true?

(a) The energy stored in a single coil having inductance L and carrying a current I is given by $W = (1/2)LI^2$.

(b) Energy storage in a circuit only takes place if the current is changing.

(c) One joule is equal to one henry–ampere^2 ($1 \; J = 1 \; H \; A^2$).

(d) A coil having a self-inductance L carries a current I and a second, similar coil is connected in series with it. The position of the coils is such that there is a mutual inductance M between them. The energy stored in the system is given by $(L + M)I^2$ joule.

(e) The energy stored in a magnetic field is proportional to its volume.

(f) The energy stored in a system of four mutually coupled circuits is given by

$$\sum_{n=1}^{4} I_n^2 R_n + \frac{1}{2}\sum_{n=1}^{4} I_n^2 L_n + \sum_{a=1,b=1,a \neq b}^{4} M_{ab} I_a I_b.$$

(g) The energy stored in a magnetic field is given by $\dfrac{B^2}{2\mu_0}$ joule per unit area.

(h) For normal values of flux density the maximum energy storage in magnetic fields is limited to about 40 J per cubic metre.

(i) The force of attraction between oppositely magnetised iron surfaces is inversely proportional to the flux density between them.

(j) The maximum available force between magnetised iron surfaces in air is of the order of 40 N m^{-2}.

(k) The force between two current-carrying coils is proportional to the product of the currents in the coils.

(l) If one of two coupled coils is rotated through an angle θ about some chosen axis, there will be a torque on it which is proportional to $dM/d\theta$ where M is the mutual inductance between the coils.

79

True: a, c, d, e, k, l

(b) So long as current is flowing in a circuit, whether alternating or steady, energy is being supplied to it. If there is resistance present some of this energy is lost in it. The rest is stored.

(f) The resistance terms do not represent stored energy. (Frame 63)

(g) It should be per unit volume, not per unit area. (Frame 65)

(h), (j) Put $B = 1$ T and $\mu_0 = 4\pi \times 10^{-7}$ in the energy and force expressions.

(i) The force is proportional to B^2. (Frame 69)

80

SUMMARY OF FRAMES 61–79

Energy stored in a magnetic field (W)

$$W = \frac{1}{2} BH \times \text{the volume of the field (J)}$$

$$W = \sum_{a=1}^{n} \frac{1}{2} L_a I_a^2 + \sum_{a=1, b=1, a \neq b}^{n} M_{ab} I_a I_b \text{ (J)}$$

Force between magnetised iron surfaces (F)

$$F = \frac{1}{2} BH \times \text{the surface area (N)}$$

Force between current-carrying circuits (F)

$$F = I_1 I_2 \frac{dM}{dx} \text{ (N)}$$

Torque on a current-carrying circuit (T)

$$T = I_1 I_2 \frac{dM}{d\theta} \text{ (N m)}$$

81

EXERCISES

(i)   A coil of 500 turns carries 10 A and is wound on a wood former 2 cm in diameter and 0.5 cm² in cross-sectional area. Calculate the energy stored.

(ii) Two coils having self-inductances of 40 μH and 60 μH are placed such that there is a coupling factor of 0.8. Calculate the stored energy when the first coil carries 5 A and the second coil carries 10 A.

(iii) The two coils of question (ii) are connected in series electrically. Calculate the two possible values of effective inductance and the corresponding energy stored when a current of 15 A is passed through them.

82

SHORT EXERCISES ON PROGRAMME 7

The numbers in brackets at the end of the questions refer to the frame where the answers may be found. All symbols are as defined in the programme.

1. State Faraday's law of electromagnetic induction. (6)
2. What is the symbol for and the unit of flux linkage? (7)
3. State Lenz's law. (8)
4. State why and describe how you would use Fleming's right-hand rule. (14)
5. Give an expression relating E, v, l and B. (14)
6. Give an expression relating I, B, F and l. (23)
7. Describe how you would use Fleming's left-hand rule. (24)
8. Give two units for induced e.m.f. (7, 8)
9. Define the ampere in terms of the force between conductors. (27)
10. Explain the meaning of 'images in infinitely permeable iron'. (28)
11. State the inductance of a coil in terms of its physical parameters. (34)
12. Give two units for self inductance. (34, 37)
13. What is the time constant of an inductive circuit? (43)
14. What is the unit of mutual inductance? (47)
15. Give the limits of the value of k (coefficient of coupling). (49)
16. Why is the 'dot notation' used? (53)
17. What is meant by the effective inductance of a circuit? (52)
18. Give an expression for the energy stored in a coil in terms L and I. (63)
19. Give an expression for the force between oppositely magnetised iron surfaces in terms of the field vectors H and B. (64)
20. Give an expression for the force between two current-carrying coils in terms of the mutual inductance between them and their currents. (73)

83

ANSWERS TO EXERCISES

Frame 20

(i)   0.9 V;
(ii) 1111;
(iii) 4.5 mWb s^{-1};
(iv) 2 mV;
(v)   107 km h^{-1};
(vi) 1.256 kV.

Frame 60

(i)   4.6 N;
(ii)   1.44 mN (attractive);
(iii) 62.5 mN (attractive);
(iv)   73.5 μH;
(v)   84 μH (assuming $k = 1$);
(vi)   337.5 μH;
(vii) 27.08 μH;
(viii) 840 nH m^{-1} (neglecting internal flux linkages);
(ix)   200 A s^{-1}, 0.25 s.

Frame 81

(i)   12.5 mJ;
(ii) 5.459 mJ;
(iii) 178.38 μH, 21.62 μH, 20.068 mJ, 2.432 mJ.

Programme 8

APPLICATION OF FARADAY'S
LAW 1 (TRANSFORMERS)

1

INTRODUCTION

In the previous programme, we saw that when two coils are placed so that they are mutually coupled, a current changing in one of them causes an e.m.f. to be induced in the other one as well as in itself, and these e.m.f.s were called 'transformer e.m.f.s'. In fact, this is the basis of the operation of the transformer which is the name given to an arrangement of coils designed to change the voltage level of electric current. It can be classified as an electrical machine, converting electric power at a high voltage, low current into electric power at a low voltage, high current or vice versa.

Transformers come in an astonishing variety of types, sizes, ratings and frequencies from a small signal transformer of a few volt amperes in electronic amplifiers to the enormous grid transformers of more than 1000 MVA used in the electricity supply industry. There are 1-phase and 3-phase transformers, 1-winding, 2-winding and 3-winding transformers, transformers for changing 3-phase to 6-phase and for changing from 3-phase to 2-phase.

In this programme we shall concentrate mainly on the 1-phase, 2-winding power transformer which operates at 50 Hz (60 Hz in the USA) and we will study:

- the operation of the ideal and the practical transformer
- the phasor diagrams associated with transformer operation
- the e.m.f. equation of the transformer
- 'equivalent' circuits of transformers
- referred quantities
- losses and efficiency of transformers
- voltage regulation
- testing of transformers
- the difference between core type and shell type transformer cores
- the difference between concentric and sandwich windings
- why 3-winding transformers are used
- why 1-winding transformers are used.

When you have studied this programme you should be able to:

- draw the phasor diagram for a transformer for any load condition
- calculate the terminal voltage of a transformer under any load condition
- predict the efficiency of a transformer at any load and power factor
- determine the condition for the maximum efficiency of a transformer and calculate its maximum efficiency at any power factor
- use the open circuit and short circuit test results to determine the equivalent circuit parameters, the iron losses and the copper losses.

2

THE SINGLE-PHASE, TWO-WINDING POWER TRANSFORMER

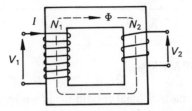

The diagram shows an iron core consisting of two 'limbs' (the upright parts) and two 'yokes' (the horizontal parts). As we shall see later, the core is usually laminated and made from special ferromagnetic material in order to reduce certain losses. A coil of N_1 turns is shown wound on the left-hand limb and a coil of N_2 turns is shown wound on the right-hand limb. In practice, each limb would have half of each coil (wound one over the other) on it in order to improve magnetic coupling. The coils are, of course, insulated from one another.

The coil connected to the supply voltage V_1 is called 'the primary winding' and the coil connected to the load is called 'the secondary winding'. When there is no load connected to the secondary winding, the transformer is said to be 'on no load' or 'on open circuit'.

3

NO LOAD CONDITIONS

Suppose that a sinusoidal voltage is applied to the primary winding with the secondary winding on open circuit. A current will flow through the primary winding and will produce a magnetic flux (Φ say). This flux will be changing with time and will link the N_1 turns of the primary winding. According to Faraday's law an e.m.f. (call it E_1) will be induced whose magnitude is given by

$$E_1 = N_1 \frac{d\Phi}{dt} \quad \text{(V)}$$

The iron core ensures that most of the flux Φ also links the N_2 turns of the secondary winding (we can assume that under ideal conditions ALL of Φ links the N_2 turns of the secondary winding). An e.m.f. will be induced in the secondary winding which we can call E_2.

Give an expression for the magnitude of E_2 and hence state the relationship between E_1, E_2, N_1 and N_2.

4

$$E_2 = N_2 \frac{d\Phi}{dt} \quad \text{(V); it follows that } \frac{E_1}{E_2} = \frac{N_1}{N_2}$$

AN IDEAL TRANSFORMER

In Frame 2 mention was made of ideal conditions in which there was perfect coupling. It is useful as a first consideration to carry this a stage further and imagine a transformer whose windings have no resistance and whose core material has a straight line $B-H$ graph which passes through the origin. This transformer will therefore:

- have no leakage inductance (no leakage flux) [leakage inductance is explained in Frame 10]
- have no $I^2 R$ loss and no IR voltage drop (no resistance)
- have no hysteresis loss (the $B-H$ graph encloses no area)
- have no eddy current loss (core material has infinite impedance).

Under these conditions, therefore, there will be no input power from the supply on open circuit, no current is required to magnetise the core and the induced e.m.f. E_1 is exactly equal to the applied voltage V_1.

5

LOAD CONDITIONS

If the secondary winding is connected to a load impedance (a sink of electrical energy) then the e.m.f. E_2 will send a current (call it I_2) around this closed circuit. The current I_2 will set up its own magnetic flux and in accordance with Lenz's law this flux will be in opposition to Φ so that it tends to oppose the changes giving rise to E_2. However, if Φ is reduced, E_1 is reduced as well as E_2 and the voltage balance between E_1 and V_1 is upset. The result is that the supply voltage sends a current around the primary circuit sufficient to restore the flux to the value necessary to induce E_1 equal in magnitude to V_1.

If the current in the secondary winding results in a current of I_1 flowing in the primary winding, what is the condition that the flux in the core remains unchanged at the original open circuit value? (Remember, flux is produced by ampere *turns*.)

6

$$\boxed{N_1 I_1 = N_2 I_2}$$

Flux is produced by the *effective* current, i.e. the current in one turn multiplied by the number of turns.

Since we are still assuming ideal conditions, there will be no voltage drop in the secondary winding on load so that the terminal voltage $V_2 = E_2$.

What then is the relationship between V_1, V_2, I_1 and I_2?

7

$$\boxed{V_1 I_1 = V_2 I_2}$$

We have seen that $N_1 I_1 = N_2 I_2$ (Frame 6) and that $N_1 E_2 = N_2 E_1$ (Frame 4).

$$\therefore \quad \frac{N_1}{N_2} = \frac{I_2}{I_1} = \frac{E_1}{E_2} \Rightarrow E_1 I_1 = E_2 I_2$$

But under ideal conditions $E_1 = V_1$ and $E_2 = V_2$, so that $V_1 I_1 = V_2 I_2$. This means that the output power is equal to the input power as we would expect of a perfect (100% efficient) machine.

We have that, under ideal conditions, $\dfrac{E_1}{E_2} = \dfrac{N_1}{N_2} = \dfrac{V_1}{V_2}$, thus $V_2 = V_1 \dfrac{N_2}{N_1}$.

The variations in the voltage at the primary winding terminals are thus reproduced at the secondary winding terminals and are magnified N_2/N_1 times.

If $N_2 > N_1$, the magnification is positive and we have a 'step-up' transformer.

If $N_2 < N_1$, the magnification is negative and we have a 'step-down' transformer.

If $N_2 = N_1$, there is no magnification (one-to-one ratio transformer).

N_1/N_2 is called the 'turns ratio' of the transformer and is often denoted by n.

A transformer having a turns ratio of 11 operates from a 240 V supply and delivers a load current of 25 A. Calculate the secondary terminal voltage and the primary winding current.

8

$$\boxed{V_2 = 21.82 \text{ V}, \; I_1 = 2.27 \text{ A}}$$

THE EQUATION FOR THE e.m.f. INDUCED IN A TRANSFORMER WINDING

Let the flux in the core be varying sinusoidally with time so that $\Phi = \Phi_m \sin(2\pi ft)$ (Wb), where Φ_m is the maximum value of the flux and f is the frequency of the supply (Hz).

Obtain expressions for the r.m.s. values of the e.m.f.s induced in the primary and secondary windings of the transformer.

9

$$\boxed{E_1 = 4.44f\Phi_m N_1 \text{ (V)}, \; E_2 = 4.44f\Phi_m N_2 \text{ (V)}}$$

Considering the primary winding,

$$e_1 = -N_1 \frac{d\Phi}{dt} = -N_1 \frac{d(\Phi_m \sin 2\pi ft)}{dt}$$

$$= -2\pi f N_1 \Phi_m \cos(2\pi ft) \text{ (V)}$$

The maximum value of this is $2\pi f N_1 \Phi_m$ (V) and the root mean square value is $\dfrac{2\pi f N_1 \Phi_m}{\sqrt{2}} = 44f N_1 \Phi_m$ (V) (see Programme 1, Frame 58).

Similarly for E_2.

Note that a sinusoidal flux wave gives rise to a cosinusoidal induced e.m.f. wave so that there is a 90° phase displacement between them.

Example

A 200 kVA, 6600/415 V, 50 Hz single-phase transformer has a primary winding of 1500 turns. Calculate the number of turns on the secondary winding, the primary winding current and the maximum value of the core flux.

Solution

The turns ratio is $6600/415 = 15.9$.

$$\therefore \quad N_1/N_2 = 15.9 \text{ and } N_2 = 1500/15.9 = 94 \text{ turns.}$$

The primary current $I_1 = 200\,000/6600 = 30.3$ A.
Using the e.m.f. equation, we have $6600 = 4.44 \times 50 \times \Phi_m \times 1500$
whence $\Phi_m = 0.0198$ Wb.

10

PRACTICAL TRANSFORMERS

A real transformer differs from the ideal transformer in that:

- Its windings will have resistance resulting in IR voltage drops and I^2R power losses in both windings.
- There will be leakage fluxes. We have discussed leakage flux in Programme 6, Frame 45, and in transformers there are leakage fluxes associated with both the primary and the secondary windings. The path of the leakage flux is mainly in air so that the reluctance is substantially constant and the flux is consequently proportional to the current. A constant inductance ($N\Phi/I$) results, giving rise to an induced e.m.f. ($L\,di/dt$) which is termed a reactance voltage. These reactance voltage drops appear in both windings.
- There will be iron losses in the form of hysteresis and eddy current losses.

The consequence of the voltage drops is that $V_1 > E_1$ and $E_2 > V_2$.
The consequence of the power losses is that $V_1 I_1 > V_2 I_2$.
The second of these indicates that the efficiency (output power divided by input power) of a transformer is less than 1 per unit (100%). Nevertheless, transformers are extremely efficient, being machines with no moving parts.

11

A PRACTICAL TRANSFORMER ON NO LOAD

When the primary winding is connected to the supply with the secondary on open circuit, a current is drawn which has two components as follows:

- A component to set up the flux in the core. This component is wholly reactive and has no power associated with it. It is 90° out of phase with the voltage. It is usually denoted by the symbol I_m.
- A component necessary to provide the hysteresis and eddy current losses in the core. This component is an active component and is in phase with the voltage. The symbol for this component is I_w.

The current drawn from the supply under these conditions is called the no load current and is denoted by the symbol I_0 and

$$I_0 = I_m + I_w \quad \text{a phasor sum}$$

This is shown in the phasor diagram in the next frame.

12

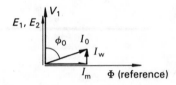

The phasor diagram of the transformer under no load conditions is shown above. The phasor representing the core flux, Φ, is taken to be the reference.

V_1 represents the supply voltage.
E_1 represents the e.m.f. induced in the primary winding.
E_2 represents the e.m.f. induced in the secondary winding.
I_0 represents the no load current.
I_w is the active component of I_0.
I_m is the reactive component of I_0 and is in phase with the flux which it produces.
ϕ_0 represents the no load phase angle and is quite large (usually $> 60°$).

Write down all the relationships between I_0, I_w, I_m and ϕ_0.

13

$$\boxed{I_w = I_0 \cos \phi_0; \; I_m = I_0 \sin \phi_0; \; I_0 = (I_w^2 + I_m^2)^{1/2}}$$

As we saw in Frame 5, when the secondary-winding induced e.m.f. sends a current I_2 through a load, there is a counterpart current produced in the primary winding. This current is equal to $I_2(N_2/N_1)$ and is given the symbol I_1. The total current flowing in the primary circuit is now the sum of the original no load current together with the primary counterpart of the secondary load current. If we let this be I_P, then

$$I_P = I_0 + I_1 \qquad \text{a phasor sum}$$

The secondary load current produces voltage drops in the resistance and leakage reactance of the secondary winding, leaving a secondary terminal voltage which is less than the secondary induced e.m.f. Similarly, the total primary current produces voltage drops in the resistance and leakage reactance of the primary winding which must be supplied by the primary supply voltage V_1 which is consequently bigger than the primary induced e.m.f.

All these points are shown in the phasor diagram of the loaded practical transformer which is given in the next frame.

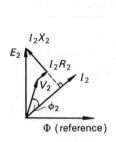

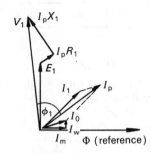

Φ is the core flux and is the reference phasor.

E_2 is the secondary induced e.m.f. and is at $90°$ to Φ as shown in Frame 8.

I_2 is the secondary load current and its phase position relative to the secondary terminal voltage depends on the quality of the load impedance. We have assumed a lagging load having a phase angle ϕ_2.

I_2R_2 is the voltage drop in the resistance of the secondary winding and is in phase with the current I_2 causing it.

I_2X_2 is the voltage drop in the leakage reactance of the secondary winding and is $90°$ ahead of the current I_2 causing it.

V_2 is the secondary terminal voltage and is $(E_2 - I_2R_2 - I_2X_2)$ phasorially.

Turning now to the primary side:

I_0, I_w and I_m are as described in the previous frame for no load conditions.

I_1 is the primary counterpart of I_2.

I_P is the total primary current and is the phasor sum of I_1 and I_0.

I_PR_P is the voltage drop in the resistance of the primary winding and is in phase with the current I_P causing it.

I_PX_P is the voltage drop in the leakage reactance of the primary and is $90°$ ahead of the current I_P causing it.

E_1 is the induced e.m.f. in the primary winding and is at $90°$ to the flux Φ.

V_1 is the supply voltage applied to the primary and is the phasor sum of E_1, I_PR_P and I_PX_P.

ϕ_1 is the phase angle of the primary circuit.

Note that the voltage drops have all been exaggerated for the sake of clarity. In practice, the resistance drops on full load might be of the order of 1 % of the respective terminal voltages and the reactance drops are about 10%. The result is that the angles between V_1 and E_1 on the primary side and between V_2 and E_2 on the secondary side are very small.

What assumption has been made about the transformer in drawing the phasor diagram?

15

> A one-to-one ratio transformer is assumed

In practice, $E_1 = nE_2$ and $I_1 = I_2/n$ where n is the turns ratio, N_1/N_2, but this would be impossible to draw on a single diagram to a reasonable scale. Phasor diagrams are therefore drawn with $N_1 = N_2$.

EQUIVALENT CIRCUITS OF TRANSFORMERS

When performing calculations, it is convenient to replace the transformer core and its windings by circuit elements which produce the same effects as the real hardware.

We begin by representing the resistance and leakage reactance of the windings by lumped parameters external to the windings themselves which may then be regarded as being perfect. This gives us the first stage in the development of the equivalent circuit and is shown below.

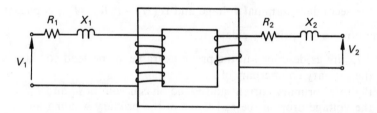

The next step is to replace the ferromagnetic core. To do this we recall that the effect of the iron is two-fold:

● it requires a magnetising current to set up a magnetic flux in it—this current is wholly reactive and lags the voltage by 90°
● it requires an active component of current to provide for its hysteresis and eddy current losses.

An inductor will draw a current which lags the voltage by 90° and therefore takes account of the magnetising current, while a resistor will draw a current in phase with the voltage and accounts for the iron losses.

The iron core may therefore be represented by a circuit consisting of a resistor and an inductor. Thus stage two of the development of an equivalent circuit takes the form shown in the next frame.

16

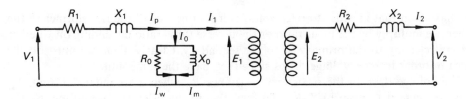

The branch in parallel with the primary winding replaces the iron core. A series combination of R_0 and X_0 could have been used but a parallel combination turns out to be more convenient.

REFERRED QUANTITIES

If we could join the two circuits together we could get rid of the windings altogether! However, if we did this there would be a big bang—unless E_1 and E_2 were equal. As it happens, we can make them equal by multiplying E_1 by N_2/N_1 or by multiplying E_2 by N_1/N_2. This effectively turns the transformer into a one-to-one ratio transformer and if we multiply E_2 by N_1/N_2 we have 'referred' E_2 to the primary.

Referred quantities are often shown dashed so that $E_2\dfrac{N_1}{N_2}$ is written E_2' and is read 'the secondary induced e.m.f. referred to the primary'.

The primary counterpart of the secondary load current is also a referred quantity because $I_1 = I_2(N_2/N_1) = I_2'$.

Write down expressions for the primary induced e.m.f. referred to the secondary, and the secondary terminal voltage referred to the primary.

17

$$E_1' = E_1(N_2/N_1); \; V_2' = V_2(N_1/N_2)$$

Before we can join the two circuits together, we must also refer the resistance and reactance of one circuit to the other. The next frame deals with this.

18

REFERRED RESISTANCE, REACTANCE AND IMPEDANCE

We have seen that in order to refer voltages from one side of the transformer to the other, we multiply by the turns ratio or the inverse turns ratio depending on whether we are referring to the primary or to the secondary. Clearly, then, we must do the same in order to refer voltage drops since they have the same unit.

The voltage drop in the secondary resistance R_2 when a secondary current, I_2, flows through it is given by $I_2 R_2$. To refer this to the primary we multiply by the turns ratio so that $I_2' R_2' = \dfrac{N_1}{N_2} I_2 R_2$.

Now $\quad I_2' = \dfrac{N_2}{N_1} I_2 \quad$ so that $\quad I_2 = \dfrac{N_1}{N_2} I_2' \quad$ and $\quad I_2' R_2' = \dfrac{N_1 N_1}{N_2 N_2} I_2' R_2$.

Finally,
$$R_2' = \left(\frac{N_1}{N_2}\right)^2 R_2.$$

Similarly,
$$X_2' = \left(\frac{N_1}{N_2}\right)^2 X_2 \quad \text{and} \quad Z_2' = \left(\frac{N_1}{N_2}\right)^2 Z_2.$$

Our equivalent circuit now becomes:

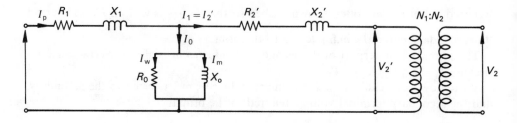

Calculations based upon this circuit, which is called 'the exact equivalent circuit referred to the primary', would yield a terminal voltage, V_2' appropriate to a 1:1 ratio transformer. To obtain the answer for any given transformer, it is simply necessary to multiply by N_2/N_1.

To emphasise this, a perfect transformer is often shown connected across the output terminals of the equivalent circuit.

It is possible to simplify the circuit considerably by moving the parallel branch to the input terminals. What are the implications of this so far as I_0 is concerned?

> If the parallel branch is at the terminals, the voltage
> across it is constant so that I_0 would be constant.

In practice, I_0 varies by about 5% as the load varies from zero to full load. However, the errors involved are very small, especially at the higher values of load, and the approximation is made in most cases.

Assuming that these relatively small errors can be tolerated, then, the approximate equivalent circuit takes the form shown below and results in much simpler calculations.

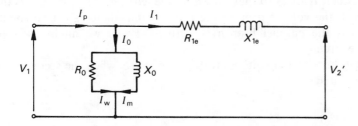

$R_{1e} = R_1 + R_2'$ and is 'the total resistance referred to the primary'.
$X_{1e} = X_1 + X_2'$ and is 'the total leakage reactance referred to the primary'.
$Z_{1e} = (R_{1e}^2 + X_{1e}^2)^{1/2}$ and is 'the total impedance referred to the primary'.

Example

A single-phase transformer has a primary winding whose resistance is 0.9 Ω and whose leakage reactance is 5.1 Ω. The corresponding values for the secondary winding are 0.01 Ω and 0.08 Ω.

Calculate the resistance of the transformer referred to the primary winding and the leakage reactance of the transformer referred to the secondary winding. The turns ratio of the transformer is 9:1.

Solution

The resistance of the secondary referred to the primary $= 0.01 \times 9^2 = 0.81$ Ω.
The total resistance referred to the primary $R_{1e} = 0.81 + 0.9 = 1.71$ Ω.

The reactance of the primary referred to the secondary $= 5.1 \times (1/9)^2 = 0.063$ Ω.
The total reactance referred to the secondary $X_{2e} = 0.063 + 0.08 = 0.143$ Ω.

20

THE NO LOAD EQUIVALENT CIRCUIT

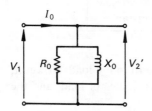 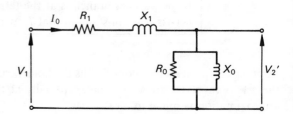

On no load or open circuit the only current flowing is I_0 so that the approximate equivalent circuit reduces to that shown on the left. The exact version is that on the right in which the voltage drop in the primary impedance Z_1 is so small that the voltage across the parallel branch is virtually V_1 so we might as well use the approximate circuit.

Now draw the 'short circuit' equivalent circuit (i.e. when the secondary terminals are short circuited).

21

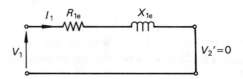

In this case the secondary induced e.m.f. is presented with an impedance which is very small so that the current in the secondary (and therefore the primary referred current) can be enormous—at least six times the full load current with normal applied voltage—so that, in comparison, I_0 is negligible. The parallel branch may therefore be omitted.

Note that on short circuit the whole of the applied voltage is used up in circulating the current through the impedance of the winding. For this reason it is referred to as the primary impedance voltage, V_{z1}

i.e.
$$V_{z1} = I_1(R_{1e} + X_{1e}) = I_1 Z_{1e} \qquad \text{phasorially}$$

22

Now decide which of the following statements are true:

(a) For all transformers $E_1/E_2 = N_1/N_2 = n$ the voltage or turns ratio.
(b) In an ideal transformer $V_1 N_1 = V_2 N_2$.
(c) Load current results in a referred primary current such that $N_1 I_1 = N_2 I_2$.
(d) The high voltage winding of a transformer is called the primary winding.
(e) The magnetising current of a transformer is in phase with the core flux.
(f) The no load power factor of a transformer is close to unity.
(g) Transformer winding induced e.m.f.s are $90°$ out of phase with the core flux.
(h) In a real transformer $V_1 > E_1$ and $V_2 > E_2$.
(i) It is possible to represent a transformer by an equivalent circuit consisting only of resistors and inductors.
(j) To refer the resistance of the primary to the secondary, multiply by $1/n^2$.

23

> True: a, c, e, g, i, j

(b) $V_1 N_2 = V_2 N_1$.   (Frame 7)
(d) The winding connected to the supply is the 'primary'.   (Frame 2)
(f) The no load current lags the applied voltage by a large angle. (Frame 12)
(h) $V_1 > E_1$ but $V_2 < E_2$.   (Frame 10)

24

SUMMARY OF FRAMES 1–23

$$\frac{E_1}{E_2} = \frac{N_1}{N_2} = \frac{I_2}{I_1} = n \quad \text{(voltage ratio = turns ratio = inverse current ratio)}$$

$E = 4.44 f \Phi_m N$ (e.m.f. equation)
$I_0 = I_w + I_m$ (no load current and its components—a phasor sum)
$I'_2 = I_2/n$ (secondary current referred to the primary)
$E'_2 = nE_2$ (secondary induced e.m.f. referred to the primary)
$Z'_2 = n^2 Z^2$ (secondary impedance referred to the primary)

25

EXERCISES

(i)   A 1-phase 6600/240 V transformer has 1500 turns on the high voltage winding. How many turns are there on the low voltage winding?

(ii) A 100 kVA 1-phase transformer with a turns ratio of 30 is connected to a 3.3 kV supply. Calculate the secondary full load current.

(iii) A 50 kVA transformer has a turns ratio of 15. The high voltage winding has 1200 turns and is connected to a 3.3 kV 50 Hz supply. Calculate the maximum value of the core flux, the secondary no load voltage and the rated primary current.

(iv) A transformer has a no load current of 1 A at a power factor of 0.4 lagging. Determine the magnitudes of I_w and I_m.

(v)   A 1-phase transformer has a turns ratio of 2 and its winding resistances are $R_1 = 0.22\ \Omega$ and $R_2 = 0.07\ \Omega$. Determine the primary resistance referred to the secondary, and the total resistance referred to the primary.

26

TRANSFORMER LOSSES

As mentioned in Frame 10, transformers are very efficient pieces of equipment. Even so, there are losses which reduce their efficiency from 100%. These are:

(a) Losses due to current flowing in the windings—the Joule effect which we discussed in Programme 4, Frame 26. Because the windings are of copper, these losses are often referred to as 'copper losses'. They are relatively small at low loads but they vary as the square of the current and become more and more significant as the load increases. The full load copper loss (i.e. when rated current is flowing) is denoted by P_c.

(b) Losses in the core of the transformer which are also called 'iron losses'. These losses, which are due to hysteresis and eddy currents, are independent of the load current and the factors upon which they depend are discussed in the following two frames. If we denote iron loss by P_i, hysteresis loss by P_h and eddy current loss by P_e, then $P_i = P_h + P_e$.

   A certain transformer has a full load copper loss of 40 W and an iron loss of 36 W. What are the respective losses at (a) half load and (b) 10% overload?

(a) 10 W and 36 W, (b) 48.4 W and 36 W

HYSTERESIS LOSS

As we saw in Programme 6, Frame 17, this may be thought of as being due to a kind of molecular friction as the atoms of the magnetic material align themselves first in one direction and then another under the influence of alternating magnetisation.

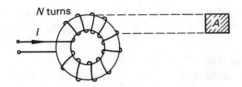

To gain a measure of the hysteresis loss, we shall consider the iron ring shown in the diagram. It has a cross-sectional area A, a mean magnetic length l, and is uniformly wound with N turns. We need to know what energy is supplied to the arrangement as the current from the supply passes through one complete cycle from 0 to I_{max} to 0 to $-I_{max}$ and back to 0 again.

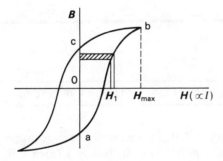

The hysteresis loop for the core material is given above.

Let a small increase in magnetising current from I_1 (corresponding to a magnetic field strength of H_1) result in an increase in flux density amounting to dB. The change in flux will be d$\Phi = A$dB and if this takes place in dt second there will be an induced e.m.f. in the coil.

Write down an expression for this induced e.m.f.

28

$$E = -N\frac{d\Phi}{dt} = -NA\frac{dB}{dt}$$

The supply voltage must have a component $V = NA\dfrac{dB}{dt}$ to provide this e.m.f.

The energy supplied in dt seconds is $VI\,dt = NA\,dBI$ (J) $= NIA\,dB$ (J).

Now $NI = H_1 l$ (Programme 6, Frame 27)

$$\therefore \qquad \text{the energy supplied} = H_1 lA\,dB = AlH_1\,dB \text{ (J)}$$

$H_1\,dB$ is the area of the shaded strip in the diagram so that the energy supplied as the current is increased from 0 to I_{max} will be proportional to the sum of all the strips such as that shown, which amounts to the total area between the B-axis and the curve ab.

Similarly the energy returned to the supply as the current is decreased to 0 again is proportional to the area enclosed by the B-axis and the curve bc.

The energy 'lost' is proportional to the difference between these two areas, i.e. the area enclosed by the B–H loop to the right of the B-axis. The same argument holds for the area to the left of the axis and the hysteresis loss as the core is taken through one complete cycle of magnetisation is given by (the area enclosed by the B–H loop) × (the volume, Al, of the core material) and is measured in joules.

Give an expression for the loss in watts if the supply frequency is f(Hz).

29

$$P_h = (\text{area of } B\text{–}H \text{ loop}) \times (\text{volume of core material}) \times f \text{ (W)}$$

It is found experimentally that the area of the loop is proportional to the maximum value of the flux density raised to some power x (area $\propto B_m^x$). The index x is called the Steinmetz index after Charles Steinmetz who first discovered the relationship. For the silicon steels used in modern transformers, x has a value between 1.8 and 2.2.

Since the volume Al is constant for a given specimen, we may write

$$P_h = k_h f B_m^x \text{ (W)} \qquad \text{where } k_h \text{ is a constant}$$

How does the hysteresis loss manifest itself?

30

As heat—the core becomes warm

EDDY CURRENT LOSS

When the alternating magnetic flux cuts the core material, it induces e.m.f. s in it and, because the core is itself a conducting material, currents will flow in it. These currents are called 'eddy currents' and produce a Joule effect, heating up the core and causing an energy loss.

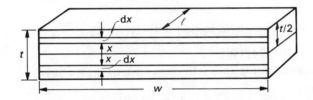

To obtain a measure of this loss, we will consider a section of the core of thickness t, width w ($\gg t$) and length l. This section may be considered as being made up of a number of rectangular blocks. Two of these are shown, the cross-section of the inner one being w wide and $2x$ thick, and the one next to it being w wide and $(2x + 2dx)$ thick. The flux is directed at $90°$ to these blocks. Let the maximum flux density in the core be B_m (T) and the frequency of the alternation be f (Hz).

Give an expression for the average e.m.f. induced in the inner rectangle.

31

$$E_{av} = 4xwB_m f \ (V)$$

Here is the reasoning: the maximum flux linking the rectangle is the maximum flux density multiplied by the area of the rectangle $= B_m \times 2xw$ (Wb). The flux changes twice per cycle and $2f$ times per second.

The r.m.s. value (E) is $\dfrac{\pi}{2\sqrt{2}} E_{av} = \dfrac{\pi}{2\sqrt{2}} 4xwB_m f = \sqrt{2}\pi xwB_m f$ (V).

What is the resistance of the path presented to this e.m.f.?

32

$$R = \frac{\rho(2w)}{(l \times dx)} \ (\Omega)$$

The resistance $R = \rho l/A$ (Programme 4, Frame 16) and in this case the length of the conductor is $2w$ (assuming that $w \gg x$) and the cross-sectional area is $dx \times l$.

The energy loss in this path is

$$\frac{E^2}{R} = \frac{2\pi^2 x^2 w^2 B_m^2 f^2}{\rho 2w/(l\,dx)}.$$

The energy loss in the whole section is therefore $P_e = \dfrac{\pi^2 w B_m^2 f^2 l}{\rho} \displaystyle\int_0^{t/2} x^2 \, dx \ (\text{W})$

$= (\pi^2 w B_m^2 f^2 l/\rho) \times (t^3/24) \ (\text{W})$.

Bringing all constants together into one constant k_e, we have

$$P_e = k_e B_m^2 f^2 \ (\text{W})$$

For a given core width (w) and length (l), how may the eddy current loss be minimised?

33

By laminating the core, making t small

B_m and f are generally fixed by the particular application.

EFFICIENCY

The efficiency of a transformer is given by $\eta = \dfrac{\text{output power}}{\text{input power}}$ per unit. The input power may be expressed as output power + losses.

$$\therefore \quad \eta = \frac{\text{output power}}{\text{output power} + \text{losses}} \ \text{p.u.}$$

If the volt-ampere rating of a transformer is S, then its output power at any fraction x of full load (if the transformer is overloaded $x > 1$) and any power factor $\cos \phi$ will be $xS \cos \phi$ (W).

If the full load copper loss is P_c, then at any fraction x of full load the copper loss will be $x^2 P_c$.

The iron loss, P_i, is constant at all loads.

The efficiency is therefore $\eta = \dfrac{xS \cos \phi}{xS \cos \phi + x^2 P_c + P_i}$ p.u.

To find the condition for maximum efficiency (η_{max}), we differentiate this expression with respect to the variable (x) and equate to zero.

Do that now and give the condition for η_{max}.

34

$$\boxed{\text{For maximum efficiency, } x^2 P_c = P_i}$$

This is obtained as follows:

$$\frac{d\eta}{dx} = \frac{xS \cos \phi (S \cos \phi + 2xP_c) - S \cos \phi (xS \cos \phi + x^2 P_c + P_i)}{(xS \cos \phi + x^2 P_c + P_i)^2}$$

Putting this equal to zero gives:

$$x(S \cos \phi + 2xP_c) = xS \cos \phi + x^2 P_c + P_i$$
$$2x^2 P_c = x^2 P_c + P_i$$
$$x^2 P_c = P_i$$
$$x = \sqrt{\left(\frac{P_i}{P_c}\right)}$$

This is independent of power factor, so the maximum efficiency always occurs at the same load regardless of the power factor.

Example

A 500 kVA, 1100/415 V, 1-phase transformer has a full load copper loss of 4 kW and an iron loss of 3 kW. Calculate:

(a) the efficiency at full load, 0.8 power factor lagging,
(b) the efficiency at three-quarters full load, unity power factor,
(c) the load at which maximum efficiency occurs,
(d) the maximum efficiency at 0.9 power factor lagging.

Solution

(a)

We have that the efficiency $\eta = \dfrac{xS \cos \phi}{xS \cos \phi + x^2 P_c + P_i}$ p.u.

At full load, $x = 1$ and for a power factor of 0.8, $\cos \phi = 0.8$

$$\text{therefore } \eta = \frac{500\,000 \times 0.8}{500\,000 \times 0.8 + 4000 + 3000} \text{ p.u.}$$

$$= 0.9828 \text{ p.u.}$$

(b)

In this case, $x = 0.75$ and $\cos \phi = 1$

$$\text{therefore } \eta = \frac{0.75 \times 500\,000}{0.75 \times 500\,000 + (0.75)^2 \times 4000 + 3000} \text{ p.u.}$$

$$= \frac{375\,000}{375\,000 + 2250 + 3000} \text{ p.u.}$$

$$= 0.9862 \text{ p.u.}$$

(c)

The maximum efficiency occurs when $x^2 P_c = P_i$, i.e. when $x = \sqrt{\left(\dfrac{P_i}{P_c}\right)}$.

For this transformer, $P_c = 4$ kW and $P_i = 3$ kW
thus the maximum efficiency occurs when $x = (3/4)^{1/2} = 0.866$
and this represents a load of $0.866 \times 500 \quad = \underline{433 \text{ kVA}}$.

(d)

The maximum efficiency when the power factor is 0.9 lagging occurs when $x = 0.866$, at which load the total losses are $2P_i = 6000$ W.

$$\eta = \frac{0.866 \times 500\,000 \times 0.9}{0.866 \times 500\,000 \times 0.9 + 6000} \text{ p.u.} = \underline{0.9848 \text{ p.u.}}$$

The *highest* efficiency obtainable with this transformer occurs at 0.866 times full load and at unity power factor

$$\text{and } \eta = \frac{0.866 \times 500\,000}{0.866 \times 500\,000 + 6000} = \underline{0.9863 \text{ p.u.}}$$

VOLTAGE REGULATION

The regulation of a transformer at some load and power factor is the difference between the secondary terminal voltage on no load and the secondary terminal voltage at the stated load and power factor. It is expressible in volts. The secondary terminal voltage on no load is the rated voltage, the secondary induced e.m.f. E_2.

For example, the full load, unity power factor regulation is given by

$$(V_2)_{\text{no load}} - (V_2)_{\text{full load}} = (E_2 - V_{2\text{full load}}) \qquad (\text{V})$$

Regulation is often expressed in per unit or percentage terms, in which case the above voltage is divided by the rated voltage (E_2) and multiplied by one (per unit) or 100 (per cent).

In practice, the variation in terminal voltage between no load and full load is very small so that the regulation is typically a few per cent.

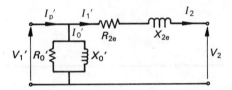 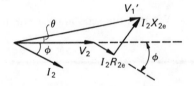

The diagram on the left above is the approximate equivalent circuit of a transformer referred to the secondary. The phasor diagram is drawn for a load current lagging by ϕ and with the voltage drops exaggerated for clarity. The angle θ is in reality very small so that

$$V'_1 = V_2 + I_2 R_{2e} \cos \phi + I_2 X_{2e} \sin \phi \qquad \text{a phasor sum}$$

On no load, $V_2 = V'_1$
therefore the regulation

$$V'_1 - V_2 = I_2 R_{2e} \cos \phi + I_2 X_{2e} \sin \phi$$

Example

A 100 kVA 1-phase transformer has a turns ratio of 5. The primary winding impedance is $(0.3 + j\, 1.1)\,\Omega$ and the secondary impedance is $(0.01 + j\, 0.035)\,\Omega$. Calculate the per unit regulation when the transformer is connected to a 2.2 kV 1-phase supply and delivers rated current at 0.8 power factor lagging.

Solution

The first step is to calculate the rated secondary current. Do that now.

37

$$\boxed{I_2 = 227 \text{ A}}$$

This is because $I_2 = \dfrac{VA \text{ rating}}{E_2} = \dfrac{100\,000}{2200/5} = 227$ A.

Next, we determine the resistance and reactance of the transformer referred to the secondary.

$R_{2e} = R_2 + R_1' = R_2 + (1/n)^2 R_1 = 0.01 + 0.3/(5)^2 = 0.022$

$X_{2e} = X_2 + X_1' = X_2 + (1/n)^2 X_1 = 0.035 + 1.1/(5)^2 \times 0.079$

The power factor is 0.8 so that $\cos \phi = 0.8$ and $\sin \phi = 0.6$, so

$$\text{the regulation} \quad = \frac{I_2 R_{2e} \cos \phi + I_2 X_{2e} \sin \phi}{E_2} \text{ p.u.}$$

$$= \frac{227 \times 0.022 \times 0.8 + 227 \times 0.079 \times 0.6}{2200/5} = \underline{0.034 \text{ p.u.}}$$

What is the secondary terminal voltage on load?

38

$$\boxed{425 \text{ V}}$$

The secondary voltage drop on full load is $0.034 E_2 = 0.034 \times 2200/5 = 15$ V, leaving a terminal voltage of $440 - 15 = 425$ V.

TESTING OF TRANSFORMERS

To determine the efficiency and regulation of a transformer at, say, full load by actually loading it and measuring the input and output power and the no load and full load terminal voltage is very expensive. Furthermore, it is not very accurate because the efficiency is obtained by dividing one large number by another large number very close to it, and because the regulation is obtained by subtracting two large numbers which are close together.

The next few frames show how some simple tests can be used to determine data which can be used to predict the efficiency and regulation performance figures at any load and power factor without actually loading the transformer.

THE OPEN CIRCUIT TEST

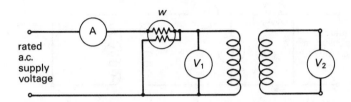

In this test, the secondary is left open circuited and the primary is supplied at rated voltage and frequency. Readings are taken of the primary voltage (V_{0c}), primary current (I_{0c}), primary power (P_{0c}) and secondary voltage (V_2).

Using these results we can obtain:

(a) the turns ratio of the transformer (V_{0c}/V_2),
(b) the no load power factor ($\cos \phi_0 = P_{0c}/V_{0c}I_{0c}$),
(c) the iron loss (P_i),
(d) the open circuit equivalent circuit parameters (R_0 *and* x_0).

(a) and (b) are straightforward.
(c) On no load the current is a few per cent of the rated current so that the power loss is less than a hundredth of the full load copper loss. The whole of the input power on open circuit may therefore be considered to be iron loss so that $P_{0c} = P_i$.

How may the results of this test be used to determine R_0 and X_0?

40

$$R_0 = \frac{V_{0c}}{I_{0c} \cos \phi_0} \quad (\Omega); \quad X_0 = \frac{V_{0c}}{I_{0c} \sin \phi_0} \quad (\Omega)$$

To satisfy yourself that this is so, refer to the equivalent circuit in Frame 20 and the phasor diagram in Frame 12.

41

THE SHORT CIRCUIT TEST

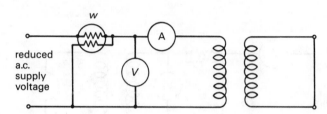

For this test, the secondary winding is short circuited and a voltage sufficient to circulate full load (rated) current is applied to the primary. Because the impedance presented to the supply voltage is very small under short circuit conditions, the applied voltage will be a small fraction of the rated primary voltage. As previously mentioned (Frame 21), if full voltage were to be applied under these conditions, very large (at least six times full load) currents would result which could damage the windings.

Readings are taken of primary voltage (V_{sc}), primary current (I_{sc}) and primary input power (P_{sc}).

Using these results:

(a) we can determine the short circuit power factor ($\cos \phi_{sc} = P_{sc}/V_{sc}I_{sc}$);
(b) since the applied voltage is very low, the flux density will be very small so that the hysteresis loss ($\propto B_{max}^x$) and the eddy current loss ($\propto B_{max}^2$) are negligibly small. The whole of the input power on short circuit may therefore be considered to be copper loss,
(c) the impedance referred to the primary is $Z_{1e} = V_{sc}/I_{sc}$ and thus the resistance R_{1e} ($= Z_{1e} \cos \phi_{sc}$) and the reactance X_{1e} ($= Z_{1e} \sin \phi_{sc}$) may be calculated.

Suppose that a short circuit test is carried out on a transformer whose rated current is 40 A but that for some reason only 35 A is circulated, resulting in a power reading of 200 W. What is the full load copper loss (P_c) of this transformer?

42

Because the test is carried out at 35 A instead of 40 A, it means that the power reading is $x^2 P_c$ (where $x = 35/40$) rather than P_c. Thus $x^2 P_c = 200$ W and $P_c = (40/35)^2 \times 200 = 261$ W.

EXAMPLE

Open and short circuit tests on a 10 kVA, 200/400 V 1-phase transformer gave the following results, all readings being taken on the 200 V side:

> Open circuit test: 200 V; 0.7 A; 60 W
> Short circuit test: 30 V; 50 A; 1100 W

Determine (a) the equivalent circuit parameters referred to the 200 V side and (b) the efficiency at three-quarters full load, 0.8 power factor lagging.

SOLUTION

(a) From the open circuit test, $\cos \phi_0 = \dfrac{W_{0c}}{V_{0c} I_{0c}} = \dfrac{60}{200 \times 0.7} = 0.428$

$$\therefore \quad \phi_0 = 64.6° \quad \text{and} \quad \sin \phi_0 = 0.903$$

$I_w = I_{0c} \cos \phi_0 = 0.7 \times 0.428 = 0.3 \text{ A} \quad \text{so that} \quad R_0 = 200/0.3 = 667 \, \Omega$

$I_m = I_{0c} \sin \phi_0 = 0.7 \times 0.903 = 0.63 \text{ A} \quad \text{so that} \quad X_0 = 200/0.63 = 317 \, \Omega.$

From the short circuit test, $\cos \phi_{sc} = \dfrac{W_{sc}}{V_{sc} I_{sc}} = \dfrac{1100}{30 \times 50} = 0.733$

$$\therefore \quad \phi_{sc} = 42.8° \quad \text{and} \quad \sin \phi_{sc} = 0.679$$

The impedance referred to the 200 V side $= Z_{1e} = V_{sc}/I_{sc} = 30/50 = 0.6 \, \Omega$
$R_{1e} = Z_{1e} \cos \phi_{sc} = 0.6 \times 0.733 = 0.44 \, \Omega$
$X_{1e} = Z_{1e} \sin \phi_{sc} = 0.6 \times 0.679 = 0.41 \, \Omega.$
  Check that the same result is obtained using $R_{1e} = W_{sc}/I_{sc}^2$ and then $X_{1e}^2 = (Z_{1e}^2 - R_{1e}^2)^{1/2}.$

(b) From the open circuit test, $P_0 = 60$ W.
  From the short circuit test, $x^2 P_c = 1100$ W.
  Now the short circuit test was carried out with a current of 50 A.
  The full load current on the 200 V side is $10\,000/200 = 50$ A.
  Thus $x = 1$ and $P_c = 1100$ W.

At three-quarters full load, then, $x = 0.75$ and $x^2 P_c = 0.56 \times 1100 = 619$ W.

At a power factor of 0.8 lagging

$$\eta = \frac{0.75 \times 10\,000 \times 0.8}{0.75 \times 10\,000 \times 0.8 + 619 + 60} \text{ p.u.}$$

$$= 0.898 \text{ p.u.}$$

Now determine the per unit regulation at full load, 0.9 power factor.

44

$$\boxed{0.1437 \text{ p. u.}}$$

Here is the working:
From the short circuit test, $R_{1e} = 0.44\ \Omega$ and $X_{1e} = 0.41\ \Omega$.
The full load regulation $= (I_{FL} R_{1e} \cos \phi + I_{FL} x_{1e} \sin \phi)$ volt.
At a power factor of 0.9 lagging, $\cos \phi = 0.9$ so that $\phi = 25.84°$
$$\text{and } \sin \phi = 0.436$$
$\therefore$ the full load regulation $= 50 \times 0.44 \times 0.9 + 50 \times 0.41 \times 0.436$
$$= 19.8 + 8.94 = 28.74 \text{ V}$$
The per unit regulation $\qquad = 28.74/200 = \underline{0.1437 \text{ p.u.}}$

45

THE SEPARATION OF THE IRON LOSS INTO ITS TWO COMPONENTS

The hysteresis loss is given by $P_h = k_h f B_m^x$ (W)
and the eddy current loss by $P_e = k_e f^2 B_m^2$ (W).
From the e.m.f. equation, the primary induced e.m.f. is given by $E_1 = 4.44 f \Phi_m N_1 = 4.44 f B_m A N_1$ where A is the cross-sectional area of the core and is a constant for a given transformer. It follows that if E_1/f can be kept constant, then the maximum flux density B_m will be constant and we may then write $P_h = k_h' f$ where $k_h' = k_h B_m^x$
$$\text{and } P_e = k_e' f^2 \text{ where } k_e' = k_e B_m^2.$$
The total iron loss $P_i = P_e + P_h$
$$= k_e' f^2 + k_h' f$$

$$\therefore \qquad \frac{P_i}{f} = k_e' f + k_h'$$

With the transformer secondary on open circuit, the primary is fed from a variable frequency, variable voltage supply. The frequency is varied in steps and at each step the voltage is adjusted so as to keep the ratio V_1/f constant, thus keeping B_m constant as explained above. Readings are taken of frequency and input power, which, because the test is carried out under open circuit conditions, is considered to be wholly iron loss (P_i). $\dfrac{P_i}{f}$ is then plotted against f.

What is the shape of this graph and what can we determine from it?

> A straight line of slope k'_e whose intercept on the vertical axis is k'_h. We can thus determine k'_e and k'_h.

Having obtained k'_e and k'_h, we can then find the two component parts of P_i at any frequency with the same value of B_m.

For a power transformer (i.e. one operating at 50 Hz) of normal design, the hysteresis loss is about four times the eddy current loss.

Example

A certain transformer has a measured iron loss of 50 W when a voltage of 125 V at 25 Hz is applied to the primary with the secondary on open circuit. At a voltage of 250 V, 50 Hz, the iron loss becomes 125 W.

Calculate the eddy current loss at 50 Hz and the hysteresis loss at 25 Hz.

Solution

The total iron loss is given by $P_i = P_e + P_h$
$$= k_e f^2 B_m^2 + k_h f B_m^x$$
where P_e is the eddy current loss, P_h is the hysteresis loss, k_e and k_h are constants, B_m is the maximum flux density and f is the frequency.

From the test data we see that V_1/f is kept constant so that B_m is constant and we can write $P_i = k'_e f^2 + k'_h f$.

Using the test results we have

$$(25)^2 k'_e + 25 k'_h = 50 \qquad \text{(i)}$$
$$(50)^2 k'_e + 50 k'_h = 125 \qquad \text{(ii)}$$

Multiplying (i) by 2 and (ii) by 1, we get

$$1250 k'_e + 50 k'_h = 100 \qquad \text{(ia)}$$
$$2500 k'_e + 50 k'_h = 125 \qquad \text{(iia)}$$

(iia)−(ia)
$$1250 k'_e \qquad\quad = 25$$
$$k'_e = 0.02$$

Substituting in (i):
$$625 \times 0.02 + 25 k'_h = 50$$
$$k'_h = 37.5/25 = 1.5$$

At 50 Hz:
$$P_e = (50)^2 \times 0.02 = \underline{50\ \text{W}}$$

At 25 Hz:
$$P_h = 25 \times 1.5 = \underline{37.5\ \text{W}}$$

What would be the total input power on no load when supplied at 300 V, 60 Hz?

47

$$\boxed{162 \text{ W}}$$

Since V_1/f is still the same, then $\boldsymbol{B}_m$ is still the same and we may use
$P_i = k'_e f^2 + k'_h f = 0.02 \times (60)^2 + 1.5 \times 60 = 72 + 90 = 162$ W.

ARRANGEMENT OF THE WINDINGS

As mentioned in Frame 2, the windings are not normally wound on separate limbs because this does not give the most effective magnetic linkage between the primary and the secondary windings. Two common arrangements are:

(a) *The core type arrangement*

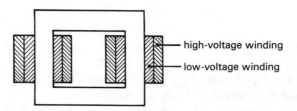

In this arrangement, half of the primary and half of the secondary windings are wound concentrically one over the other on each limb.

(b) *The shell type arrangement*

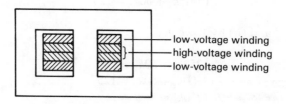

The windings in this type are usually wound side by side in a sandwich arrangement as shown in the diagram.

Although the shell type is mechanically stronger, the core type is easier to repair on site and is more often used in the UK and in Europe.

Why do you think the windings are placed in the relative positions shown (i.e. with the low voltage winding nearer the core)?

48

> It is easier to insulate between the core and the low voltage winding than between the core and the high voltage winding.

In addition, there is a higher current in the low voltage winding so that the forces of repulsion between its turns are greater. It is mechanically stronger if the high voltage winding is wrapped around it.

THREE-WINDING TRANSFORMERS

A 1-phase transformer may have 3 windings, the third one being referred to as the 'tertiary'. This may be needed to supply a load at a different voltage from that on the secondary winding or which, though requiring the same voltage as that on the secondary, must be kept separate from it. There are other reasons for using tertiary windings in 3-phase transformers to do with assisting in unbalanced loads and reducing harmonics, but these are beyond the scope of this programme.

Just as the current in the secondary winding has a counterpart in the primary, so too will the tertiary winding load current have its counterpart in the primary. The total current in the primary winding of a 3-winding transformer is thus given by

$$I_P = I_0 + I'_2 + I'_3 \qquad \text{phasorially}$$

How do you read I'_3?

49

> 'the tertiary current referred to the primary'

Example

A 3-winding 1-phase transformer is supplied at 11 kV and is loaded as follows:

secondary (400 V): 200 kVA at a power factor of 0.9 lagging
tertiary (110 V):   75 kVA at a power factor of 0.6 leading

The no load current is 5.5 A, 0.156 power factor lagging. Calculate the primary and current factor.

The solution is given in the next frame, but try it yourself, first.

50

The secondary load power is given by $V_2 I_2 \cos \phi_2$

$$\therefore \quad I_2 = \frac{200\,000}{400 \times 0.9} = 555 \text{ A} \quad \text{and} \quad I_2' = I_2 \frac{N_2}{N_1} = 555 \frac{400}{11\,000} = 20.2 \text{ A}$$

In complex form: $I_2' = 20.2(\cos \phi_2 - j \sin \phi_2)$ A where $\phi_2 = \cos^{-1} 0.9$

$$\therefore \quad I_2' = 20.2(0.9 - j\,0.436) \text{ A}$$

Now you determine the tertiary current referred to the primary in complex form.

53

$$\boxed{I_3' = 11.36(0.6 + j\,0.8) \text{ A}}$$

You should have found that the tertiary current is 1136 A ($75\,000/(110 \times 0.6)$). The j-term is positive because the current is leading.

We now have

$$
\begin{aligned}
I_0 &= 5.5(0.156 - j0.988) = (0.858 - j5.434)\text{ A}\\
I_2' &= 20.2(0.9 - j0.436) = (18.18 - j8.81)\text{ A}\\
I_3' &= 11.36(0.6 + j0.8) = (\underline{6.82 + j9.09})\text{ A}\\
&\therefore \quad I_P = (25.86 - j5.154)\text{ A}
\end{aligned}
$$

$I_P = (25.86)^2 + (5.154)^2)^{1/2} = \underline{26.37 \text{ A}}$

the power factor $= \cos \phi_1$ lagging where $\phi_1 = \tan^{-1}(5.154/25.86)$

$\phi_1 = 11.27°$ and $\cos \phi_1 = \underline{0.98 \text{ lagging}}$.

52

1-WINDING TRANSFORMERS

These are also known as autotransformers and are used in order to economise on conductor material. In a three-phase form they are commonly used for interconnecting the 132 kV grid system to the 400 kV supergrid, and ratings in excess of 1000 MV A are in operation.

We shall consider the 1-phase autotransformer, the arrangement of which is shown in the next frame, in order to see how a saving in copper is effected.

What determines the amount of conductor material required in a transformer?

The amount of current and the number of turns required

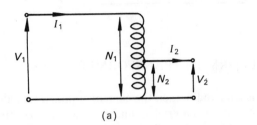

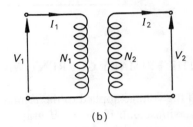

(a)   (b)

The 1-phase autotransformer arrangement is shown in diagram (a) above and the equivalent 2-winding transformer is given in diagram (b).

The primary consists of N_1 turns and the load is supplied from the e.m.f. developed across the common section of N_2 turns. The secondary m.m.f. opposes the primary m.m.f. so the current in the common section is $I_2 - I_1$ which is less than the current I_2 required in the secondary of the equivalent 2-winding transformer. The total ampere turns on the 2-winding transformer is given by $N_1 I_1 + N_2 I_2 = 2N_1 I_1$ since $N_1 I_1 = N_2 I_2$.

Obtain an expression for the total ampere turns on the autotransformer.

54

$$2N_1 I_1 - 2N_2 I_1$$

This comes about because there is a current of I_1 ampere in the $(N_1 - N_2)$ turns of the top section, and a current of $(I_2 - I_1)$ ampere in the N_2 turns of the common section. The total ampere turns are thus $I_1(N_1 - N_2) + (I_2 - I_1)N_2$, and since $N_1 I_1 = N_2 I_2$, this reduces to the expression in the box.

$$\frac{\text{Amount of conductor material in circuit (a)}}{\text{Amount of conductor material in circuit (b)}} = \frac{2N_1 I_1 - 2N_2 I_2}{2N_1 I_1} = 1 - \frac{N_2}{N_1}$$

The saving is greater as $N_2 \to N_1$, and it is found that if N_1 is more than about three times N_2, there is not much advantage in using an autotransformer.

Can you see any disadvantages of 1-winding transformers?

55

> If an open circuit were to occur in the common section, the primary voltage would be applied to the load.

THE PRIMARY CURRENT WAVEFORM IN TRANSFORMERS

If the voltage applied to the primary is sinusoidal, then the current supplied will also be sinusoidal. If the $B-H$ graph of the core material is linear over the operating region, then the flux–magnetising current graph will also be linear and the flux waveform will be the same as that of the current. However, in order to minimise the volume of core material used, modern transformers operate with high values of maximum flux density so that the non-linear portion of the $B-H$ graph is encountered. As the current reaches its maximum values during the cycle, therefore, the flux tends to level off giving a flat-topped waveform containing many odd harmonics, the third harmonic being the most prominent. Because $E \propto d\Phi/dt$, the e.m.f. waveform is also non-sinusoidal and contains a significant amount of third harmonic. The harmonics of the induced e.m.f. are not balanced by similar components in the applied voltage, so they circulate harmonic currents around the primary which add to the original sinusoidal current to produce a current waveform with harmonics in it. The transformer arranges things so that this current is now of the correct shape to produce a sinusoidal flux wave which in turn produces an e.m.f. waveform which is sinusoidal in shape.

Now answer true or false to the following statements:

(a) Iron losses in transformers are independent of load current.
(b) Copper losses in transformers are proportional to f^2.
(c) Hysteresis loss is proportional to the area of the hysteresis loop.
(d) Iron losses are minimised by laminating the core.
(e) The maximum efficiency of a transformer occurs when $x = (P_c/P_i)^{1/2}$.
(f) The per unit regulation at full load is given by $(V_{n1} - V_{f1})$.
(g) The voltage regulation of a transformer is commonly greater than 90%.
(h) An open circuit test on a transformer enables its iron loss to be found.
(i) Short circuit tests are performed with normal voltage and frequency applied to the primary.
(j) In core type transformers, half of each winding is wound concentrically on each limb.
(k) Shell type transformers usually have sandwich type windings.
(l) Autotransformers are used to economise on core material.

> True: a, c, d, h, j, k

(b) $P_c \propto I^2$. (Frame 26)

(e) Maximum efficiency occurs when $x = (P_i/P_c)^{1/2}$. (Frame 34)

(f) This must be divided by V_{n1} to obtain the per unit regulation. (Frame 36)

(g) Regulation is usually quite small ($<20\%$). (Frame 36)

(i) This test is carried out with a very much reduced voltage. (Frame 41)

(l) They are used to economise on conductor material. (Frame 52)

57

SUMMARY OF FRAMES 26–56

Transformer copper loss is a variable loss and varies as the square of the load current. The full load copper loss is represented by P_c (W) and at any fraction, x, of the full load, the copper loss is given by $x^2 P_c$ (W).

Transformer iron loss (P_i) is a constant or fixed loss and comprises two parts:
Eddy current loss $P_e = k_e f^2 B_m^2$
Hysteresis loss $P_h = k_h f B_m^x$ where x is known as the Steinmetz index.

The efficiency of a transformer is given by $\eta = \dfrac{xS \cos \phi}{xS \cos \phi + x^2 P_c + P_i}$ p.u. where x is the fraction of the full load current, S is the rating of the transformer in VA and $\cos \phi$ is the power factor of the load.

The maximum efficiency occurs when the variable loss equals the fixed loss i.e. when $x^2 P_c = P_i$ and $x = \sqrt{\dfrac{P_i}{P_c}}$.

The voltage regulation of a transformer is given by $(V_{n1} - V_{f1})$ (V). The per unit regulation is given by $\dfrac{V_{n1} - V_{f1}}{V_{n1}}$ p.u.

$$\frac{\text{conductor material required in an autotransformer}}{\text{conductor material required in the equivalent 2-winding transformer}} = 1 - \frac{N_2}{N_1}$$

58

EXERCISES

(i) A 100 kVA 1-phase transformer has a full load copper loss of 1 kW and an iron loss of 800 W. Calculate:

(a) the copper loss at one-quarter and three-quarters full load,
(b) the efficiency at half load, 0.8 power factor lagging,
(c) the efficiency at full load, unity power factor,
(d) the fraction of full load at which the maximum efficiency occurs.

(ii) Calculate the maximum efficiency of the transformer of question (i) when it operates with a power factor of 0.8.

(iii) What is the highest possible efficiency at which the transformer of question (i) can operate and under what conditions?

(iv) A 125 kVA, 2000/400 V 1-phase transformer gave the following test results on the high voltage side:

open circuit: 2000 V; 1 A; 1000 W
short circuit: 65 V; 40 A;   750 W

Calculate (a) the approximate equivalent circuit parameters,
(b) the efficiency at full load, 0.8 power factor,
(c) the per unit regulation at full load, 0.6 power factor lagging.

59

SHORT EXERCISES ON PROGRAMME 8

The numbers in brackets at the end of each question refer to the frames where the answers are to be found. All symbols have the meaning given in the programme.

1. State the relationship between E_1, E_2, N_1 and N_2. (4)
2. Give the relationship between N_1, N_2, I_1 and I_2. (6)
3. Explain what happens to the primary current of a transformer whenever its secondary current increases. (5)
4. What is meant by a 'step-up' transformer? (7)
5. Write down an expression for the induced e.m.f. in the primary winding of a transformer in terms of B_m, f, A and N. (9)
6. State why the no load current of a transformer is not purely reactive. (10)
7. Draw the exact equivalent of a transformer referred to the primary. (18)
8. Draw the approximate equivalent circuit referred to the primary. (19)

9. What must be done to refer the primary resistance of a transformer to the secondary? (18)
10. How does P_h vary with frequency and with maximum flux density? (29)
11. How does P_c vary with load? (26)
12. Define the voltage regulation of a transformer. (36)
13. State the condition for the maximum efficiency of a transformer. (34)
14. How may P_e be minimised? (33)
15. What is the Steinmetz index? (29)
16. Describe two common transformer core constructions. (47)
17. Describe two common transformer winding arrangements. (47)
18. State why a transformer may have a tertiary winding. (48)
19. Explain why autotransformers are used. (52)
20. What can be determined from the open- and short-circuit tests? (39, 41)

60

ANSWERS TO EXERCISES

Frame 25

(i)   55;
(ii) 909 A;
(iii) 12 mWb, 220 V, 15.15 A;
(iv) 0.4 A, 0.917 A;
(v)   0.055, 0.5.

Frame 58

(i)   (a) 62.5 W, 562.5 W, (b) 0.9744 p.u., (c) 0.9823 p.u., (d) 0.894;
(ii) 0.978 p.u.;
(iii) 0.9842 p.u. when operating at unity power factor and with a load of 0.894 × 100 kVA = 89.4 kVA;
(iv) (a) $R_0 = 4000\ \Omega$, $X_0 = 2309\ \Omega$, $R_{1e} = 0.468\ \Omega$, $X_{1e} = 1.556\ \Omega$, (b) 0.9725 p.u., (c) 0.0476 p.u.

Programme 9

APPLICATION OF FARADAY'S LAW 2 (ROTATING MACHINES)

1

INTRODUCTION

The operation of all rotating electromagnetic machines is based upon two principles which we have already discussed in Programme 7. These are:

(i) the electromagnetic induction of electromotive force in a conductor or circuit due to changing magnetic flux linking it (Frame 6);

(ii) the development of a force or a torque on a conductor which is carrying current and which is situated in a magnetic field (Frames 23, 73 and 74).

Although all machines rely for their operation on these two principles, there are different 'types' of machine, broadly classified as follows:

> the direct current (d.c.) machine,
> the three-phase alternating-current (a.c.) induction machine,
> the three-phase a.c. synchronous machine,
> and the single-phase a.c. machine.

All electromagnetic machines are energy convertors. We saw in the previous programme that transformers are machines which convert electrical energy at some level of voltage and current into electrical energy at some other voltage and current. Rotating electromagnetic machines can convert electrical energy into mechanical energy (in which case they are called motors), or mechanical energy into electrical energy (in which case they are called generators).

Direct current machines are most often used in the motoring mode and because they have excellent torque and speed range capabilities, they are used extensively in industry for driving rolling mills and other machinery requiring these qualities.

Three-phase induction machines are also most often used as motors. They are relatively cheap and easy to maintain and they account for the vast majority of the world's industrial motors.

Three-phase synchronous machines are widely used in both modes and in the generating mode they supply most of the power generated by the supply authorities in England and Wales.

Single-phase machines are mainly used as motors and are made in relatively small sizes. They are widely used for driving domestic equipment.

In this programme we will consider the principle of operation of all of these types of rotating machine. In the case of the d.c. motor and the three-phase induction motor we will study the performance characteristics in more detail.

2

A SIMPLE ELECTROMAGNETIC MACHINE

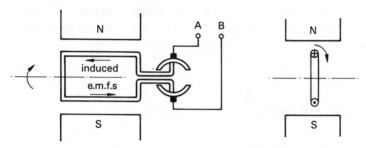

The diagram shows a single coil rotating in a steady magnetic field produced by a permanent magnet (the field could equally be produced by a current-carrying coil wound around pieces of iron known as pole pieces). Since the coil sides are cutting magnetic flux there will be e.m.f.s induced in them, and these e.m.f.s will be alternating as the coil sides come under the influence of first the North pole and then the South pole. The alternating e.m.f.s are connected to a device called a commutator which is mounted on the rotating part of the machine and which acts to reverse the connections to the external circuit at the instant at which the coil sides move from the influence of one polarity to the other. It acts as a rectifier, in fact, so that the external circuit is presented with e.m.f.s of one polarity only. The connection between the commutator and the external circuit is made by means of carbon 'brushes'.

Is the machine shown a motor or a generator?

3

> A generator because the direction of motion is opposite to the direction of the electromagnetic force

The e.m.f. obtained from the simple d.c. generator is unidirectional but is not constant. In order to obtain a substantially constant e.m.f. a large number of coils are used and are connected to different segments of a multi-segment commutator in such a way that at successive instants the carbon brushes are connected to a coil which is undergoing the same rate of change of flux. By moving the brushes around the commutator the e.m.f. presented to the external circuit can be varied from zero to a maximum.

How may the machine shown above be changed into a motor?

4

> By removing any load from the terminals AB and connecting a
> d.c. supply across them. Electrical energy is then fed into the
> machine which is allowed to rotate in the direction in which
> it is urged by the electromagnetic force

THE CONSTRUCTION OF A d.c. MACHINE

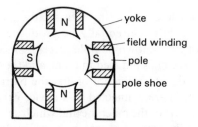

The field system shown consists of coils wound around pole cores which project
inwards from a cylindrical frame known as the yoke. The purpose of the field system
is to produce a magnetic field in the air gap of the machine. In d.c. machines the field
system is always mounted on the stationary part of the machine which is called the
stator. A 4-pole field system is shown in the diagram above.

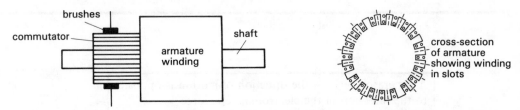

The *armature* is the name given to the winding which carries the main current
whose magnetomotive force (m.m.f.) interacts with the field m.m.f. to produce torque.
In d.c. machines the armature winding is always mounted on the rotating part of the
machine which is called the rotor.

The rotor consists of a laminated steel core, cylindrical in shape, and the conductors
of the armature winding are located in slots around its inner periphery. The
commutator, which consists of a number of copper segments each insulated from the
one next to it, is mounted on the rotor.

Why is the rotor laminated?

5

> To reduce eddy currents which are produced by the e.m.f.s induced by the relative motion between the rotor conductors and the magnetic flux.

TYPES OF ARMATURE WINDING

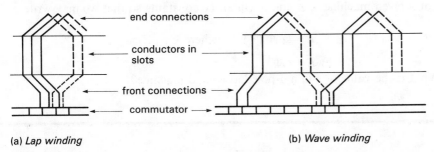

(a) *Lap winding*   (b) *Wave winding*

The diagram on the left shows a 'lap' winding in which the two sides of the same coil are connected to adjacent commutator segments but are placed in slots such that they are approximately one pole pitch apart. Their e.m.f.s are therefore acting in the same direction around the coil. It can be shown that using this type of winding there are as many parallel paths through the armature (from positive brush to negative brush) as there are poles.

The diagram on the right shows a 'wave' winding in which the two sides of a coil are not on adjacent segments but are also placed in slots such that they are approximately a pole pitch apart. It can be shown that with this type of winding there are just two parallel paths through the armature regardless of the number of poles.

6

THE e.m.f. INDUCED IN A d.c. ARMATURE

A machine having $2p$ poles is rotating at ω radians per second (rad/s). The time taken for a coil side to move through 1 pole pitch is $(2\pi/2p\omega)$ (s) so that if the flux per pole is Φ weber, the e.m.f. induced in the coil side is given by

$$\frac{d\Phi}{dt} = \frac{\Phi}{(2\pi/2p\omega)} = \frac{p\omega\Phi}{\pi} \quad \text{(V)}.$$

Suppose that in total there are Z conductors making up the armature winding and that there are c parallel paths through the armature, what will be the total e.m.f. induced and measurable between the brushes?

7

$$p\omega\Phi Z/\pi c \quad (\text{V})$$

This is because there are Z/c conductors in series in each of the c parallel paths so that the total e.m.f. is the e.m.f. per conductor multiplied by the number of conductors in series in each path.

For a given machine p, Z and c will all be constants so that we may write

$$E = k\omega\Phi \ (\text{V}) \qquad \text{where } k = pZ/\pi c$$

or $E \propto \omega\Phi$

What is the value of c in a 4-pole lap wound machine?

8

$$c = \text{the number of poles} = 4$$

THE TORQUE DEVELOPED IN A d.c. ARMATURE

If E is the e.m.f. induced in the armature and I_a is the current flowing through it, then the power in the armature is EI_a (W) and this will be

- leaving the armature if the machine is generating, or
- entering the armature if the machine is motoring.

Neglecting losses, this power must equal the mechanical power which is

- supplied to the armature if the machine is generating or
- supplied by the armature if the machine is motoring.

If the torque acting is T (N m) and the speed of the rotor is ω (rad s^{-1}) then the mechanical power is given by ωT (W).
Thus
$$\omega T = EI_a$$

and
$$T = \frac{EI_a}{\omega} \quad (\text{N m})$$

By substituting for E and using k to represent all the constants in any given machine, obtain a simplified expression for the torque.

$$\boxed{T = k\Phi I_a \qquad \text{(N m)}}$$

Substituting for $E = p\omega\Phi Z/\pi c$ into the expression for torque gives

$$T = \frac{p\omega\Phi Z I_a}{\pi c\omega} = \frac{p\Phi Z I_a}{\pi c} \qquad \text{(N m)}$$

For a given machine p, Z and c are constants so we may write

$$T = k\Phi I_a \qquad \text{where } k = pz/\pi c$$

We may also write $\qquad T \propto \Phi I_a$

Example

A d.c. machine has 8 poles each of which has a useful flux of 80 mWb. The armature winding has 576 conductors and the rotor speed is 500 r/min.
(a) Calculate the induced e.m.f. if the armature is lap wound.
(b) Determine the torque developed and the power in the armature when each conductor carries its rated current of 50 A and the e.m.f. is that calculated in part (a).

Solution

(a) $\qquad E = \dfrac{p\omega\Phi Z}{\pi c} = \dfrac{4 \times (2\pi \times 500/60) \times 0.08 \times 576}{\pi \times 8} = \underline{384 \text{ V}}$

Note that the speed in rad s^{-1} = $2\pi \times$ the speed in r s^{-1}.

(b) $\qquad T = \dfrac{p\Phi Z I_a}{\pi c} = \dfrac{4 \times 0.08 \times 576 \times (50 \times 8)}{\pi \times 8} = \underline{2934 \text{ N m}}$

Note that there are 8 parallel paths through the armature so that the total armature current is $8 \times 50 = 400$ A.
p is the number of pole *pairs* $= 8/2 = 4$.
(c) The armature power is $EI_a = 384 \times 400 = \underline{153.6 \text{ kW}}$
    What would be the armature power if it were rewound as a wave winding?

10

$$\boxed{153.6 \text{ kW as before}}$$

If the armature is wave wound there are just 2 parallel paths through the armature so that $c = 2$. This means that the induced e.m.f. is 4 times bigger than before. However, the armature current is now only $2 \times 50 = 100$ A which is $1/4$ of the previous value and the product EI_a is unchanged.

COMMUTATION

This is the name given to the reversal of the e.m.f. and current in any armature coil which takes place when the commutator segments to which it is connected are short circuited by a brush.

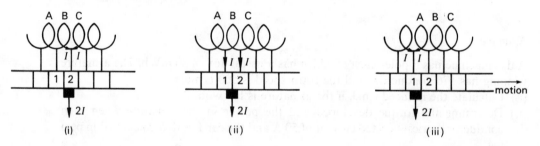

(i)   (ii)   (iii)

In diagram (i) coil B is on the point of being short-circuited and it carries half of the brush current. In diagram (ii) commutator segments 1 and 2 are short-circuited, thus short-circuiting coil B which then carries no current. Diagram (iii) shows conditions immediately after the short circuit, and coil B should then carry the current I (half the brush current) in the opposite direction from that in diagram (i).

If the current in coil B has not reached its full value, I, in the new direction by the time the commutator segment 2 has left the brush, then, since coil C *is* carrying the full current I, the difference must flow to the brush via the commutator segment 2 and this takes place in the form of an arc.

The reversal of the current has to take place in a very short time (of the order of milliseconds) and e.m.f.s are induced in the coil being commutated as a result of its self-inductance and the mutual inductance between it and the adjacent coils. These e.m.f.s constitute what is known as a reactance voltage and act to oppose the reversal of the current. Because inductance depends on the number of turns on a coil, single turn coils are used wherever possible.

Other methods of improving commutation, that is of reducing the arcing which damages the commutator and the brushes and introduces losses are:

(a) The use of interpoles

Interpoles, which are also known as compoles, are narrow poles situated mid-way between the main poles as shown in diagram (a) below.

In this method an e.m.f. is induced in the short-circuited coil in such a direction as to neutralise the reactance voltage. This e.m.f. must therefore be in the same direction as that of the current after commutation has taken place. The flux which induces this e.m.f. must be of the same polarity as the next main pole in the direction of rotation in the case of a generator, but the next main pole behind in the case of a motor. The interpoles are connected in series with the armature.

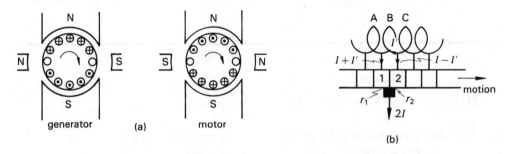

generator   (a)   motor

(b)

(b) The use of high resistance contact brushes

Diagram (b) shows coil B undergoing commutation. If the brushes are of low resistance $I' \to 0$ and remains so until the brush has left segment 2. However if the brush is of high contact resistance, then r_1 increases and r_2 reduces as the commutator moves to the right. I' thus increases throughout the commutating period until, at the end of it, $I' = I$.

The use of high resistance brushes together with interpoles has made possible virtually sparkless commutation from no load to 1.25 times full load.

Now decide which of the following statements are true.

(a) There is no fundamental difference between the construction of a d.c. motor and a d.c. generator.
(b) In a d.c. machine the armature winding is usually on the rotor.
(c) The field winding of a d.c. machine is wound on a cylindrical core.
(d) Wave wound armature windings are used where high currents are required.
(e) For a given d.c. machine $E \propto \omega$.
(f) For a given d.c. machine, $T \propto I_a \Phi$.
(g) In a 2-pole machine $c = 2$, regardless of whether the armature is wave wound or lap wound.
(h) Interpoles are always of the opposite polarity to the next main pole in the direction of rotation.

11

> True: a, b, f, g

(c) The field coils are on salient poles. (Frame 4)

(d) Except in 2-pole machines, lap windings have more parallel paths through the armature and are thus better suited to high current machines. (Frame 5)

(e) $E \propto \omega \Phi$. (Frame 7)

(h) Not in the case of a generator. (Frame 10)

12

SUMMARY OF FRAMES 1–11

The e.m.f. generated in a d.c. armature: $\quad E = p\Phi\omega Z/\pi c$ (V)

For a given machine, $\quad E \propto \omega\Phi$

The torque developed in a d.c. armature $\quad T = p\Phi Z I_a/\pi c$ (N m)

For a given machine, $\quad T \propto \Phi I_a$

For a lap wound machine $\quad c = 2p$ where $p =$ the number of pole pairs.

For a wave wound machine $c = 2$.

13

EXERCISES

(i) A 4-pole d.c. machine has a lap wound armature with a total of 564 conductors. The flux per pole is 30 mWb and the machine is driven at a speed of 1000 r/min. Calculate (a) the e.m.f. generated and (b) the torque developed when the armature current is 30 A.

(ii) A wave wound armature has 410 conductors and runs at 1200 r/min. If the field system has 4 poles and the e.m.f. generated is 500 V, calculate the flux per pole.

(iii) A d.c. machine runs at 800 r/min, the e.m.f. generated is 150 V and the armature current is 20 A. With the flux unchanged and the speed increased to 1700 r/min, the torque increases by 50%. Calculate the new values of induced e.m.f. and armature current.

14

d.c. MOTORS

The vast majority of d.c. machines operate as motors and in this mode:

(i)   electrical energy is converted into mechanical energy,
(ii) all losses are supplied in the form of electrical energy,
(iii) the rotor must provide both the load torque and the friction torque,
(iv) the voltage at the machine terminals must provide the induced e.m.f. and the voltage drops in the machine.

There are two main parts to the d.c. motor: the armature system which is mounted on the rotor and the field system which is mounted on the stator. The armature winding, which is of copper, will have resistance (call it r_a) and inductance (L_a). It will also have an e.m.f. (E) induced in it as it rotates in the magnetic field produced by the field system, and this generated e.m.f. will be constant so long as the speed of rotation is constant. The armature circuit may therefore be represented by an equivalent circuit as shown in diagram (a). V is the supply voltage and I_a is the armature current.

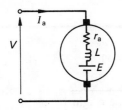

(a)

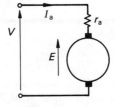

(b)

The inductance may be neglected under steady state operating conditions and then the equivalent circuit is usually drawn as shown in diagram (b).

Write down Kirchhoff's voltage equation for the armature circuit.

15

$$\boxed{\begin{array}{ll} V = E + I_a r_a + L_a(di_a/dt) & \text{circuit (a)} \\ V = E + I_a r_a & \text{circuit (b)} \end{array}}$$

The field system consists of coils wound around pole pieces and connected so as to provide alternate North and South poles. D.C. motors are classified according to the way in which the field system is supplied and this may be:

(a) separately from the armature (separately excited motors),
(b) from the same supply as the armature (series, shunt and compound motors).

16

SEPARATELY EXCITED MOTORS

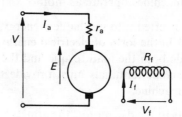

The equivalent circuit of the motor under steady state conditions is as shown in the diagram. Again, inductance effects are absent. V_f, I_f and R_f are respectively the field supply voltage, the field current and the field resistance. Whereas the armature winding resistance r_a is very small ($< 1 \, \Omega$) the field winding resistance is of the order of $10^2 \, \Omega$.

Can you think what would be the shape of a graph of torque to a base of load current (I_a) if friction is neglected? Refer to Frame 9 for a hint!

17

A straight line through the origin

This is because $T \propto \Phi I_a$ and in this case Φ is constant (because I_f is constant) so that $T \propto I_a$.

SHUNT EXCITED MOTORS

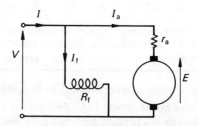

In this type of motor the field winding is connected in parallel (shunt) with the armature. As in the case of the separately excited field winding, it has a high resistance and carries a current, I_f, of the order of 1 A. There are two independent Kirchhoff voltage equations:

$$V = E + I_a r_a$$

and $$V = I_f R_f$$

There is also one Kirchhoff current equation. Write it down.

18

$$\boxed{I = I_a + I_f \text{ (A)}}$$

Again the torque–load current graph is a straight line since the flux is constant. The torque characteristic of a shunt motor is therefore similar to that of a separately excited motor and is shown below.

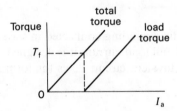

T_f is the torque required on no load to overcome mechanical losses (friction).
What can you say about the e.m.f. of shunt and separately excited motors?

19

$$\boxed{\text{Since } E = k\omega\Phi \text{ and } \Phi \text{ is constant, then } E \propto \omega}$$

Example

A d.c. shunt motor has an armature resistance of 0.4 Ω, a shunt field resistance of 160 Ω and is fed from a constant d.c. supply of 200 V. It runs at a speed of 850 r/min and takes an armature current of 50 A. Calculate (a) the field current and (b) the generated e.m.f. when the speed is halved.
The solution is given in the next frame. Try it yourself, first.

20

Solution

(a) $I_f = V/R_f = 200/160 = \underline{1.25 \text{ A}}$

(b) Initially $V = E_1 + I_a r_a \Rightarrow E_1 = V - I_a r_a$

$$\therefore \qquad E_1 = 200 - 50 \times 0.4 = 180 \text{ V}$$

We have that $E \propto \omega\Phi$ and in this case Φ is constant since the field current is constant so that $E \propto \omega$.

This means that $\dfrac{E_1}{E_2} = \dfrac{\omega_1}{\omega_2}$ and if the speed is halved, $\omega_2 = \omega_1/2$

$$\therefore \qquad E_2 = E_1/2 = \underline{90 \text{ V}}$$

21

Series motors have their field winding connected in series with the armature winding and because it carries the full load current it is designed to have a very low resistance (R_s). The steady state equivalent circuit takes the form shown in the diagram.

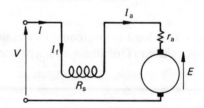

Clearly, there is only one current involved so the armature current, the field current and the supply current are all one and the same. The Kirchhoff voltage equation is $V = E + I_a(r_a + R_s)$. The torque–load characteristic is obtained as follows: $T \propto \Phi I_a$ and in this case I_a is also the field current which produces the field flux so that, so long as the *B–H* graph for the iron is linear, $\Phi \propto I_a$ and we have $T \propto I_a^2$.

However, as the load current I_a becomes large, the field current becomes large and the *B–H* graph begins to show signs of saturation. After saturation has occurred, Φ is constant so that $T \propto I_a$.

We have, then, $\qquad\qquad T \propto I_a^2 \qquad$ before saturation

and $\qquad\qquad\qquad\qquad T \propto I_a \qquad$ after saturation

The torque characteristic therefore takes the following form:

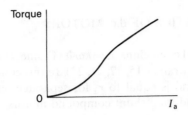

So far as the e.m.f. is concerned, we can say that $E \propto \omega$ ... true or false?

> False. Φ is not constant so we must write $E \propto \omega\Phi$

COMPOUND EXCITED MOTORS

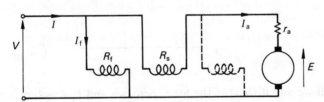

A compound wound motor has a series field and a shunt field and by appropriate choice of the number of turns on each, almost any characteristic between a series and a shunt characteristic may be obtained. The shunt field may be connected on the supply side of the series field in which case it is termed a 'long shunt', or on the armature side of the field winding (shown dotted), which is called a 'short shunt'.

The series field and the shunt field may be wound so that the fluxes they produce are in the same sense and this is called 'cumulative compounding'. If on the other hand they are wound so that their fields are in opposite directions, then this is termed 'differential compounding'.

For the long shunt connection, the Kirchhoff equations are

$$V = E + I_a(r_a + R_s)$$
$$V = I_f R_f$$
$$I = I_a + I_f$$

23

SPEED CHARACTERISTICS OF d.c. MOTORS

We have seen that for a d.c. machine $E = k\omega\Phi$ (Frame 7) and that in general for a d.c. motor $V = E + I_a R_a$ (Frames 15, 17, 21, 22). In this equation, R_a is the resistance of the armature circuit and is equal to r_a for separately excited and shunt motors and $(r_a + R_s)$ for series and long shunt compound motors.

Rearranging these two equations we have $\omega = \dfrac{E}{k\Phi}$ and $E = V - I_a R_a$.

Substituting for E in the expression for speed gives $\omega = \dfrac{V - I_a R_a}{k\Phi}$.

We can now deduce the speed characteristic for the various types of motor. Considering first the separately excited motor and the shunt motor, in both cases the flux is constant so that $\omega \propto V - I_a R_a$. If we neglect $I_a R_a$ as being small compared with V, then $\omega \propto V$ and therefore with a constant supply voltage the speed is constant. This is shown as graph (i) in the diagram.

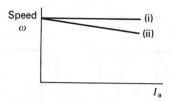

However if we admit that, as the load increases and I_a gets larger, the term $I_a R_a$ is no longer negligible, then the speed characteristic will fall as shown in graph (ii). Even at full load, the term $I_a R_a$ is very small compared with V, so that the drop in speed is not great.

These motors are used for general-purpose constant-speed applications which are not subject to large overloads. However as we shall see later, if speed variation is required, it is easy to arrange.

In the case of the series motor, the flux is proportional to the field current, neglecting saturation so that $\Phi \propto I_a$, since I_a is the field current.

$$\therefore \quad \omega \propto \frac{V - I_a R_a}{I_a}$$

Neglecting $I_a R_a$ and with a constant supply voltage V we have $\omega \propto 1/I_a$.

What can you deduce about the no load speed of a series motor?

It is very high

The graph of ω/I_a takes the form shown from which we see that as $I_a \to 0$ so $\omega \to \infty$. For this reason, a series motor should never be used where it might become disconnected from its load (for example, with a belt drive which could break). Series motors are used for varying loads where constant speed is not necessary (for example, for cranes and for traction applications).

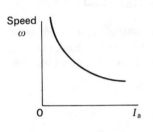

(a) *Series motor*

(b) *Compound motor*

Compound motors normally have shunt characteristics which are modified by a weak series field winding. A differentially compounded machine could be used to offset the speed drop on large loads due to the $I_a R_a$ term in the speed equation. However, with this type of compounding, unstable operation is possible, particularly at overloads, and it is possible that the series field could take over at starting and cause the machine to rotate in the wrong direction. Differential compounding is not common, therefore!

Cumulative compounded motors are useful for fluctuating loads such as rolling mills and for hoists and goods lifts. The shunt field winding ensures that the speed cannot reach dangerously high levels under low load operating conditions.

25

SPEED CONTROL OF d.c. MOTORS

We saw in Frame 23 that the speed of a d.c. motor is given by $\omega = \dfrac{V - I_a R_a}{k\Phi}$

from which we see that for a given I_a, ω may be varied by:

(i)   varying the resistance of the armature circuit, thus varying $I_a R_a$,
(ii) varying the resistance of the field circuit, thus varying Φ,
(iii) varying the voltage applied to the armature, i.e. varying V.

We shall consider each of these methods in the following frames.

26

(i) Variation of armature circuit resistance

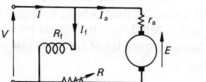 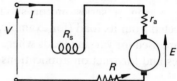

This is achieved by placing a variable resistor R, called a controller, in series with the armature. The armature circuit resistance is thus increased which means that the speed can only be reduced below the nominal (rated) value using this method.

Can you think of any disadvantages of this method?

27

> It is very wasteful of energy because the full armature current flows through the controller causing high $I^2 R$ losses. Because the controller has to be continuously rated, it is expensive.

This method is only used where efficiency is not a prime consideration or where small speed drops for short intervals of time are required. Examples are cranes and traction. The graphs show how, at a given load, the speed falls as R is increased. Note also that the shunt characteristic becomes much steeper as R is increased because of the large $I_a R$ voltage drop.

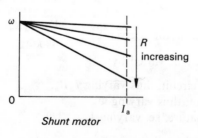 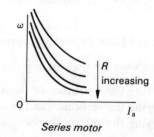

Shunt motor   *Series motor*

Example

A d.c. shunt motor having an armature resistance of 0.4 Ω takes an armature current of 25 A from a 220 V supply. It runs at 1000 r/min.

Calculate the resistance required to be placed in series with the armature in order to reduce the speed to 500 r/min, if the load torque is (a) constant and (b) proportional to the square of the speed.

Solution

When the speed is 1000 r/min, $E_1 = V - I_{a1}r_a = 220 - 10 = 210$ V.
When the speed is 500 r/min, E will change and I_a could change depending on the load torque requirements.

$\dfrac{E_1}{E_2} = \dfrac{\omega_1 \Phi_1}{\omega_2 \Phi_2}$ and, since Φ is constant for a shunt motor, $E_2 = \dfrac{\omega_2 E_1}{\omega_1}$.

At 500 r/min, $\omega_2 = \omega_1/2$ so that $E_2 = E_1/2 = 105$ V.
We have that $T \propto I_a \Phi$ but, because Φ is constant in this case, $T \propto I_a$.

(a) For a constant load torque the armature current is constant so that at 500 r/min
$I_a = 25$ A.
Now $E_2 = V - I_{a2}(r_a + R)$
$\qquad 105 = 220 - 25(0.4 + R)$
and $\underline{R = 4.2\ \Omega}$

(b) For a load torque which is proportional to the square of the speed, since $T \propto I_a$ and $T \propto \omega_2$ then $I_a \propto \omega^2$

i.e. $\qquad \dfrac{I_{a2}}{I_{a1}} = \dfrac{\omega_2^2}{\omega_1^2} = (1/2)^2 \qquad$ and $\qquad I_{a2} = I_{a1}/4 = 25/4 = 6.25$ A

Thus $105 = 220 - 6.25(0.4 + R)$
and $\underline{R = 18\ \Omega}$

29

(ii) Variation of field circuit resistance

This is achieved by placing a variable resistor, called a field regulator, in series with the field winding in the case of shunt and compound excited motors. In series motors it is necessary to divert some of the armature current away from the field winding by placing a resistor in parallel with it.

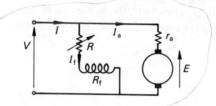 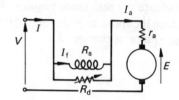

In this method the field current can only be reduced so that the flux is reduced and since this appears in the denominator of the speed expression, the speed can only be increased above the nominal.

What do you think limits the speed range?

30

The torque available. Since $T \propto \Phi$, the torque at high speeds is very much reduced because Φ is then very small.

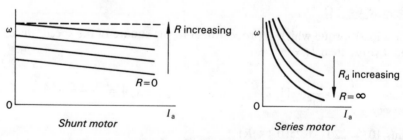

Shunt motor   *Series motor*

Speeds of up to 4 times the nominal are obtainable. The method is not wasteful of energy because the current through the field regulator is quite small. The graphs show how the speed at a given load varies with the regulator resistance and the divertor resistance.

31

Example

A d.c. series motor has an armature resistance of 0.2 Ω, a field resistance of 0.15 Ω and takes 50 A from a 220 V supply when running at 750 r/min. Calculate the speed at which the motor will run when a divertor resistance of 0.15 Ω is placed in parallel with the field winding. Assume that (i) the flux is proportional to the field current and (ii) the armature current remains the same.

Solution

At 750 r/min, $E_1 = V - I_a(r_a + R_s) = 220 - 50(0.2 + 0.15) = 202.5$ V. With the divertor resistance in circuit, the total armature circuit resistance becomes $r_a + \dfrac{R_s R_d}{R_s + R_d} =$ 0.275 Ω and $E_2 = 220 - 50 \times 0.275 = 206.25$ V.

With the divertor resistance connected, the field current and therefore the flux will be half of its previous value, i.e. $\Phi_2 = \Phi_1/2$.

Now $E \propto \omega\Phi$ which means that $\dfrac{E_1}{E_2} = \dfrac{\omega_1 \Phi_1}{\omega_2 \Phi_2}$ and $\omega_2 = \dfrac{E_2 \omega_1 \Phi_1}{E_1 \Phi_2}$.

If N is the speed in r/min, then $N = 60\omega/2\pi$ and if $E \propto \omega\Phi$, then $E \propto N\Phi$.

It follows that $N_2 = \dfrac{E_2 N_1 \Phi_1}{E_1 \Phi_2}$ r/min

i.e.
$$N_2 = \frac{206.25 \times 750 \times \Phi_1}{202.5 \times \Phi_1/2} = \underline{1528 \text{ r/min}}$$

(iii) Variation of the voltage applied to the armature

By varying the voltage applied to the armature, i.e. the term V in the expression $\omega = \dfrac{V - I_a R_a}{k\Phi}$, the speed of a motor may be varied from zero to several times the nominal speed. Furthermore if the polarity of V is reversible, the motor can be made to run in the 'forward' and the 'reverse' directions. We shall now consider how this may be achieved.

33

WARD LEONARD CONTROL

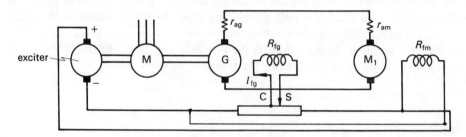

In this method, the variable voltage is obtained from a separate motor–generator (M–G) set. The diagram shows the basic arrangement in which a 3-phase motor drives a d.c. generator at constant speed, Any type of constant-speed motor can be used as the M–G set drive motor, though the most commonly used is the 3-phase induction motor. The field of the motor whose speed is to be controlled (M_1 in the diagram) is constantly energised.

How do you think the variable voltage (V) is obtained for supplying to M_1?

34

> By varying the field current of the d.c. generator G

If the slider S is moved to the right or to the left of the centre (C) of the generator field regulator, the field current I_{fg} of the d.c. generator may be varied from zero to maximum in either the positive or negative direction. The flux (Φ) in the generator will therefore be variable from zero to maximum in either polarity and so its induced e.m.f. E ($\propto \omega\Phi$ with ω constant) will be variable from zero to maximum either positively or negatively. This e.m.f., less the voltage drop in the armature resistance of the generator, is supplied to the motor whose speed is to be controlled (M_1) as V, and since the field of this motor is constantly energised, its speed is variable from zero to maximum in either direction.

The disadvantages of this method are the high initial cost of the M–G set and the high maintenance costs.

POWER ELECTRONIC CONTROL OF d.c. MOTORS

In the late 1950s/early 1960s, grid-controlled mercury arc rectifiers were introduced as an alternative to the M–G set part of the Ward Leonard system. These were multi-anode, single cathode (a pool of mercury) devices fed from a 3-phase supply via a 3-phase to 6-phase transformer. The instants in the a.c. cycle at which conduction commenced was controllable by means of voltages applied to 'grids' placed between anodes and cathode. The mean d.c. output voltage from the rectifier and supplied to the motor was thus variable.

More recently, solid state rectifiers using thyristors have superseded the mercury arc rectifiers. They are now widely used and form the basis of what has become known as power electronics.

There are two common power electronic methods for controlling the voltage applied to the armature of a d.c. motor from an a.c. supply. Both methods use thyristors as a rectifying device or as a switch. A thyristor is a three-terminal device which will conduct whenever its anode is more positive than its cathode provided that there is a suitable signal at its third terminal (known as the gate). Applying a signal to the gate is referred to as 'firing' or 'triggering' and once a thyristor has been fired it will continue to conduct until its cathode becomes more positive than its anode or until the current falls below a certain low value called the 'holding current'.

(i) Phase control

In this method, the fraction of a cycle for which the (rectified) a.c. supply voltage is transmitted to the motor terminals is controlled by delaying the instant at which the thyristor is fired. The mean value of the d.c. voltage applied to the motor armature is therefore variable.

This method is used in equipments ranging from 1-phase, single thyristor circuits for controlling hand tools and domestic machinery to 3-phase bridge circuits containing six or twelve thyristors controlling motors driving rolling mills and other machinery requiring upwards of 1 MW.

(ii) Integral cycle control

In this method, which is also known as burst firing, the supply voltage is transmitted to the motor for a number of complete cycles and then the firing pulses are removed from the thyristors so that for a number of complete cycles no voltage is allowed to reach the motor terminals. By varying the ratio of the number of complete cycles 'on' to the number of complete cycles 'off', the average value of the voltage applied to the motor armature is varied and its speed is controlled.

This method is only suitable for small motors (less than 1 kW).

36

Chopper control

In this method the motor whose speed is to be controlled is connected to a d.c. supply via a circuit which switches the d.c. supply to the motor at rapid and regular intervals. The mean voltage applied to the motor is variable by varying the ratio of 'on' to 'off' time. In this case, once fired, the thyristors have to be forced off at appropriate times and this is accomplished by connecting capacitors, charged in the correct sense across the anode and cathode, so that the thyristors are reverse biassed.

This method is used for controlling traction motors on railways and for electric road vehicles using battery supplies.

37

STARTING OF d.c. MOTORS

At standstill there is no generated e.m.f. in a d.c. armature so that in the equation $V = E - I_a r_a$, $E = 0$ which means that $I_a = V/r_a$.

Now the supply voltage is several hundred volts and r_a is typically less than 1 Ω so that, if the motor were switched directly on to the supply, the armature current would tend to become more than 1000 A. This could cause damage to the motor. In practice, therefore, a variable resistance is connected in series with the armature during the starting period and is reduced as the motor speed and therefore E increases from zero.

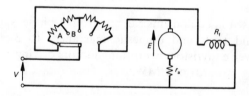

With the starter in position A, the current I_a rises to the maximum allowable $I_{max} = (V - E)/(r_a + R)$, R being the resistance of the starter. As the motor gathers speed, E increases and I_a will fall until it reaches some value I_{min}. At this point the starter is moved to position B and I_a rises to I_{max} again because the resistance has been reduced. I_a falls again as E increases further. The process of reducing resistance continues until all the starting resistance has been removed and I_a and E are at their normal values.

What determines the design values of I_{max} and I_{min}?

38

> I_{max} must not exceed a safe value; I_{min} must be large enough for the motor to accelerate against the load torque.

Now decide which of the following statements are true.

(a) In d.c. motors, all losses are supplied in the form of mechanical energy.
(b) The armature winding of a d.c. motor has a very low resistance.
(c) The field winding of all d.c. motors is of the order of 100 Ω.
(d) Armature windings of d.c. motors have no inductance.
(e) The field winding of d.c. motors is located on the stator.
(f) The purpose of the field winding is to produce alternate N and S poles.
(g) The torque characteristic of a shunt motor is a square law.
(h) For shunt and separately excited motors, $E \propto \omega$.
(i) The speed of a series motor falls slightly on load.
(j) The speed of a shunt motor may be increased by weakening the field.
(k) The speed of a separately excited motor can be varied over a wide range in either direction using the Ward Leonard system of voltage control.
(l) Speed control of d.c. motors by power electronic means can only be achieved if there is an a.c. supply available for rectification.
(m) Burst firing is another name for integral cycle control.
(n) Shunt motors require starting resistance to keep I_a to a safe level.

39

> True: b, e, f, h, j, k, m, n

(a) Losses are supplied in the form of electrical energy.   (Frame 14)
(c) Series motors have a low field winding resistance.   (Frame 21)
(d) All coils have inductance.   (Frame 14)
(g) It is linear over the normal load range.   (Frame 18)
(i) For a series motor, $\omega \propto 1/I_a$.   (Frame 23)
(l) Chopper control can be used with d.c. supplies.   (Frame 36)

40

SUMMARY OF FRAMES 14–39

For a d.c. motor $V = E + I_a R_a$ where R_a is the total resistance in the armature circuit.
For any motor $E \propto \omega \Phi$
For a shunt and a separately excited motor $E \propto \omega$ since Φ is constant.
For any motor $T \propto \Phi I_a$.
For a shunt and a separately excited motor $T \propto I_a$ since Φ is constant.
For a series motor $\qquad T \propto I_a^2 \qquad$ before saturation
$\qquad\qquad\qquad\qquad T \propto I_a \qquad$ after saturation

The nominal speed of a d.c. motor is given by $\omega = \dfrac{V - I_a R_a}{k\Phi}$ rad s^{-1}

The speed may be varied by variation of R_a, $\Phi(I_f)$ or V.

41

EXERCISES

(i)   A d.c. shunt motor draws 30 A from a 200 V d.c. supply. Its armature and field resistances are respectively 0.4 Ω and 100 Ω. Calculate the field current and the generated e.m.f.

(ii) A 220 V d.c. shunt motor having an armature resistance of 0.4 Ω takes 22 A from the supply when driving a certain load. Its field resistance is 110 Ω. What resistance must be placed in series with the armature to halve the speed if the load torque is constant.

(iii) A shunt motor takes an armature current of 20 A from a 240 V d.c. supply when running at 800 r/min. Its armature resistance is 0.4 Ω while that of the field winding is 160 Ω. Calculate the resistance which must be placed in series with the field in order to increase the speed to 900 r/min. Assume the flux to be proportional to field current.

(iv) A 230 V series motor runs at 500 r/min when its armature current is 25 A. Its armature and field windings each have a resistance of 0.4 Ω.
  (a) Determine the speed when the current is 20 A, assuming the flux to be proportional to the field current.
  (b) If a 0.4 Ω resistor is placed in parallel with the field and the armature current is 25 A, calculate the speed at which the motor runs assuming the flux to be 75% of its previous value.

42

d.c. GENERATORS

d.c. supplies are now commonly obtained from a.c. supplies by means of rectification so that d.c. generators have become less important and demand for them has declined over recent years. They are still manufactured, however, in a wide range of sizes from a few watts to megawatts.

In the generating mode of operation:

(i)   all losses must be supplied in the form of mechanical energy,
(ii) the driving motor must provide the generator and the friction torque,
(iii) the generated voltage must provide all voltage drops.

Generators are classified by the manner of their excitation, which is the same as for motors, and we shall now consider these in turn.

SEPARATELY EXCITED GENERATORS

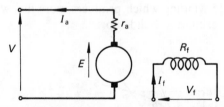

The steady state equivalent circuit is shown in which all the symbols have the meanings defined in the case of a separately excited motor.

Write down Kirchhoff's voltage law equation for the armature circuit.

43

$$V = E - I_a r_a \ldots \text{ the graph of which is given below}$$

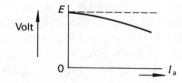

This indicates that the terminal voltage will fall as the load current increases. In fact, V falls by up to 10% from no load to full load.

44

SHUNT EXCITED GENERATOR

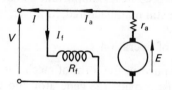

With the generator being driven at constant speed, an e.m.f. will be generated in the armature if there is a magnetic field for its conductors to cut. In this self-excited generator the current producing the field is derived from the e.m.f. generated in the armature. There seems to be a snag here! How can the machine work?

Well, by relying, initially, on remanent flux density in the poles. An e.m.f. will be induced in the armature conductors as they cut the residual flux. This e.m.f. sends a current through the field winding which produces more flux which induces a larger e.m.f., and so on as the excitation builds up.

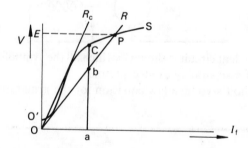

As the field current increases the armature voltage increases along O'S which is called the open circuit characteristics. There will be a 'family' of these characteristics, each one corresponding to a different speed of rotation. The line OR has a slope E/I_f ($= R_f$, neglecting r_a in comparison with R_f).

At any given value of field current, say Oa, ac is the e.m.f. generated and ab is the voltage across the field resistance. The remainder, bc, is available to increase the field current. At point P, stability is reached and OE represents the open circuit voltage of the generator when it is driven at the given speed if its field winding resistance is R_f.

What would be the effect if the field resistance were such that the slope of OR were greater than that of OR_c?

> The build up of field current would not be possible because bc (see the diagram) is zero. R_c is called the critical field resistance.

For a given field resistance R_f, there will be a critical speed below which the machine will not generate. This is the speed corresponding to the open circuit characteristic whose initial part has a slope of OR_f.

The load characteristic of a shunt generator over the normal operating range is similar to that for a separately excited generator in that it falls from $V = E$ on no load in accordance with $V = E - I_a r_a$ as the load current rises.

Example

The open circuit characteristic of a 400 V d.c. generator when driven at a speed of 700 r/min is given by:

Generated e.m.f. (V):   120   240   334   400   444   470
Field current (A):   0.5   1.0   1.5   2.0   2.5   3.0

The field resistance is 160 Ω and the armature resistance is 0.5 Ω.

Determine: (a) the open circuit voltage of the generator,
(b) the critical resistance at a speed of 700 r/min,
(c) the terminal voltage when the generator supplies 50 A.

Hints for solution

Plot the open circuit characteristic using the given data.

Superimpose the field resistance line using the origin as one point and the point (2.5 A, 400 V) as the other.

(a) The resistance line intersects the open circuit characteristic at the point where $E = 465$ V.

(b) To find the critical resistance, draw a line tangential to the initial part of the open circuit characteristic. The slope of this line gives the critical resistance. You should obtain $R_c = 250$ Ω.

(c) The terminal voltage $V = E - I_a r_a$. When the armature content is 50 A, $V = 465 - 50 \times 0.5 = 440$ V.

Series generators have load characteristics similar to the open circuit characteristic because the field current and the armature (load) currents are one and the same. This type of generator is therefore of little use for generating (even approximately) constant voltage.

Compound generators have both a shunt field winding and a series field winding, just as in the case of motors. As we have seen, the shunt generator terminal voltage falls as the load current is increased from zero to the rated value. In the compound wound generator, the series field may assist the shunt field such that the terminal voltage on load is the same as on no load, in which case the generator is said to be level compounded. If the fall in voltage is reduced but not eliminated, the machine is undercompounded and if the terminal voltage on full load is actually higher than it is on no load, the machine is over compounded. All of these characteristics are shown in the diagram together with the differentially compounded characteristic which is the result of arranging for the series field to oppose the shunt field.

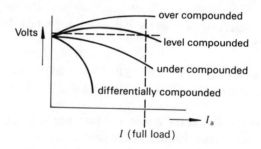

Example

A 400 V, 80 kW d.c. generator has a shunt field winding of 1200 turns per pole. The shunt field current required to give the rated voltage on no load is 4.1 A while on full load the current required is 5.8 A.

Calculate the number of series field turns per pole if the machine is to operate as a level compounded generator.

Solution

Shunt field ampere turns per pole on full load $= 5.8 \times 1200 = 6960$.
Shunt field ampere turns per pole on no load $= 4.1 \times 1200 = 4920$.
Ampere turns per pole required from the series field $= 2040$.

Full load armature current $= 80\,000/400 = 200$ A.

$\therefore$ number of turns per pole $= 2040/200 = 10.2$ say 10.

47

ARMATURE REACTION

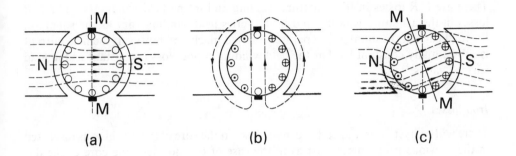

(a)   (b)   (c)

When no armature currents are flowing, the flux in the air gap is due to the field system alone as shown in diagram (a) in which MM is called the magnetic neutral axis and is at right angles to the flux lines. When armature currents flow, they set up their own m.m.f. as shown in diagram (b) in which the field flux is ignored. Diagram (c) shows the resultant flux due to the fluxes in (a) and (b) acting together. It can be seen that the magnetic neutral axis has been shifted.

This distorting effect of the armature current is called armature reaction.

Assuming anticlockwise rotation, is the machine shown a motor or a generator?

48

> A generator because it is running in the opposite direction
> to that in which it is urged by the electromagnetic force on it.

In the case of a motor the magnetic neutral axis would be shifted against the direction of rotation. Sketch diagrams to prove this for yourself.

The effect of armature reaction is to cause the flux density to be stronger on one side of the pole face and weaker on the other. If the poles are not saturated, the *total* flux is not changed but it is redistributed. However, under normal conditions and given the relatively high values of magnetic flux density used in modern machines, saturation of the poles will occur. There is then, when the armature current is at its rated value, a reduction in the air gap flux of up to about 8%.

It is very difficult to take this into account in calculations because the effect is non-linear, but fortunately it can be neglected without introducing very great errors in most cases.

49

LOSSES AND EFFICIENCY OF d.c. MACHINES

Copper losses

There are $I^2 R$ losses in the armature winding and in the field winding(s). The $I^2 R$ losses in the armature winding and in the series field winding vary as the square of the load current. In shunt and separately excited machines, however, the field current is substantially constant so that the field winding copper loss is constant as the load current varies.

Iron losses

There will be hysteresis and eddy current losses in the core of the machine as discussed in the previous programme. Just as in the case of transformers, the core of the d.c. machine is laminated in order to reduce eddy current losses and hysteresis loss is minimised by the appropriate choice of the quality of the steel.

Mechanical losses

These are made up of bearing friction loss, which is approximately proportional to the speed of the rotor, and windage loss, which is proportional to the square of the peripheral speed of the armature (or the cube of the speed if a fan is fitted).

Commutator/brush loss

There is an 'electrical' loss due to the contact resistance between the commutator segments and the brushes. It is usual to assume a constant voltage drop of up to 1 V per brush arm and this produces a loss of $V_b I_a$ where V_b is the assumed brush drop and I_a is the armature load current.

There is also a 'mechanical' loss due to the friction between commutator and brush which depends on brush pressure, the coefficient of friction for the brush material and the speed of rotation.

These losses are small compared with those in the other groups and are often neglected in calculations where great accuracy is not required.

Efficiency

The efficiency of a d.c. machine is given by $\eta = \dfrac{\text{output power}}{\text{input power}} \text{p.u.}$

Efficiency can also be expressed in terms of machine losses as follows:

$$\eta = \frac{\text{output power}}{\text{output power} + \text{losses}} \text{p.u.} \qquad \text{or} \qquad \eta = \frac{\text{input power} - \text{losses}}{\text{input power}} \text{p.u.}$$

MOTOR EFFICIENCY

For a d.c. motor, the input power (P_{in}) is the terminal voltage (V) multiplied by the current (I) fed into the motor. The output power (P_{out}) is obtained by subtracting the various losses so that

$$P_{out} = P_{in} - I_a^2 R_a - I_f^2 R_f - P_{fw} - P_i$$

where P_{fw} is the friction and windage loss, P_i is the iron loss, $I_a^2 R_a$ is the copper loss in the armature circuit and $I_f^2 R_f$ ($= V_f I_f$) is the copper loss in the field circuit.

The motor efficiency,

$$\eta_m = \frac{VI - I_a^2 R_a - I_f^2 R_f - P_{fw} - P_i}{VI} \text{p.u.}$$

For a shunt motor, $I = I_a + I_f$ and under normal (full load) conditions $I_a \gg I_f$ so that $I \approx I_a$.

Putting $I = I_a$ and dividing throughout by I_a we get

$$\eta_m = \frac{V - I_a R_a - (I_f^2 R_f + P_{fw} + P_i)/I_a}{V} \text{p.u.}$$

The terminal voltage V is substantially constant so that the efficiency will be maximum when the numerator is maximum. To find the condition for maximum efficiency, then, we differentiate the numerator with respect to the variable (I_a) and equate to zero:

$$\frac{d}{dI_a}(V - I_a R_a - (I_f^2 R_f + P_{fw} + P_i)/I_a) = -R_a - (-I_a^2)(I_f^2 R_f - P_{fw} - P_i) \quad \text{(i)}$$

Equating to zero gives $\quad I_a^2 R_a = (I_f^2 R_f + P_{fw} + P_i)$.

Differentiating expression (i) again gives a negative result, indicating a maximum value for η_m.

Using the same symbols, obtain an expression for the efficiency (η_g) of a d.c. generator.

$$\eta_g = \frac{VI}{VI + I_a^2 R_a + I_f^2 R_f + P_{fw} + P_i} \text{p.u.}$$

In the case of a generator, V is the output voltage and I is the output current. The input power is therefore VI plus the losses.

52

PREDICTION OF EFFICIENCY

The efficiency of a machine can be determined by direct measurement of input and output power on a load test. In the case of a motor the load could be a brake of some kind while for a generator the output power could be used to supply a resistance. These methods are wasteful of energy and for large machines are not only inadvisable but are often not possible.

A more convenient and cheaper method is a so-called indirect method. This consists of measuring the losses and then predicting the efficiency at any load using the equations for η_m or η_g. The most common indirect test is the Swinburne test in which the machine is run as a motor on no load and at normal voltage and speed.

Because the voltage is the normal value so too will be the iron losses.
Because the speed is normal so too will be the friction and windage losses.
Because the armature current is small (the motor is on no load) the copper loss in the armature winding will be negligible. The whole of the input power to the armature can therefore be considered to be the normal iron, friction and windage loss (the 'fixed' loss).

The armature winding and the field winding resistances are then measured and are adjusted to take account of the normal operating temperature. For any value of load current I, the efficiency can now be predicted.

Example

A 400 V d.c. shunt motor, when subjected to a Swinburne test took 4 A from the supply. The armature and field winding resistances, adjusted to normal operating temperature, are 0.2 Ω and 200 Ω respectively.

Predict the efficiency of the motor when running on full load and taking a current of 100 A.

Solution

Field current $= 400/200 = 2$ A.
On no load, input power to the armature $= 400 \times (4 - 2) = 800$ W.
This is considered to be the iron, friction and windage loss.
At full load, the input power to the motor is $400 \times 100 = 40\,000$ W.
The armature copper loss at full load $= (100 - 2)^2 \times 0.2 = 1921$ W.
The field circuit loss $= 400 \times 2 = 800$ W.
The iron friction and windage loss $= 800$ W.
Total loss at full load $= 1921 + 800 + 800 = 3521$ W.
Output power at full load $= 40\,000 - 3521 = 36\,479$ W.
Efficiency $= 36\,479/40\,000 = \underline{0.912}$ p.u.

53

Now decide which of the following statements are true.

(a) The terminal voltage of a separately excited d.c. generator falls sharply as the load current increases.
(b) A shunt generator relies on residual magnetism for its operation.
(c) A d.c. generator will not generate if its field resistance is less than a certain critical value.
(d) For a given field resistance there is a critical speed below which a d.c. generator will not generate.
(e) A series generator has a constant load characteristic.
(f) Compound generators are usually wound cumulatively.
(g) For a level compounded generator, V (no load) = V (full load).
(h) The voltage drop on full load of an under compounded generator is greater than that of a shunt generator of the same rating.
(i) Armature reaction in d.c. machines distorts the field flux only if the poles are saturated, otherwise there is no effect.
(j) The 'fixed' losses of a d.c. machine can be found from a no load test.

54

True: b, d, f, g, j

(a) The terminal voltage falls slightly on load. (Frame 43)
(c) R_f must be $< R_c$. (Frame 45)
(e) The terminal voltage varies widely with load current. (Frame 46)
(h) The voltage drop is less than that of a shunt machine. (Frame 46)
(i) The flux is reduced when there is saturation. (Frames 47, 48)

55

SUMMARY OF FRAMES 42–54

For a d.c. generator, $V = E - I_a R_a$ where R_a is the total resistance in the armature circuit.

$$\eta_g = \frac{\text{output power } (VI)}{\text{output power} + \text{losses}} \text{p.u.}; \qquad \eta_m = \frac{\text{input power} - \text{losses}}{\text{input power } (VI)} \text{p.u.}$$

η_{max} occurs when variable loss $(I_a^2 R_a)$ = fixed loss $(I_f^2 R_f + P_{fw} + P_i)$.

56

TEST EXERCISE

(i) A d.c. shunt generator gave the following open circuit characteristic when driven at 300 r/min:

Field current (A):	0	2	3	4	5	6	7
Terminal voltage (V):	7.5	93	135	165	186	202	215

Determine: (a) the open circuit voltage if the field resistance is 40 Ω,
(b) the critical field resistance at 300 r/min.

(ii) The shunt winding of a 50 kW, 250 V d.c. generator has 1000 turns per pole and requires 5 A to produce rated voltage at full load current. On no load the field current required for this voltage is 3.5 A.

Calculate the number of series turns per pole for level compounding.

(iii) The total input power to a d.c. shunt motor when supplied with its rated voltage of 460 V and run on no load was 3220 W. The resistance of the armature circuit is 0.2 Ω and that of the field circuit is 230 Ω.

Determine: (a) the efficiency when the motor takes its full load current of 130 A,
(b) the maximum possible efficiency of the motor and the armature current at which it occurs.

57

3-PHASE a.c. MACHINES

These machines rely for their operation on the production of a rotating magnetic field which is one whose magnitude is constant but whose direction rotates in space in contrast to an alternating field whose magnitude varies positively or negatively along a fixed axis.

A rotating magnetic field is produced in a simple 2-pole machine as follows: three-phase currents are fed into three windings placed mutually at 120° on the stator of the machine as shown in the diagram (a) on the next page. The coil sides of winding 1 are 1 & 1′ while the corresponding coil sides of the other two windings are 2 & 2′ and 3 & 3′.

Diagram (b) shows the waveform of the three-phase currents which are fed into the corresponding ends of the three windings. This diagram shows that:

(i) at instant t_1, $i_1 = 0$, i_2 is positive and i_3 is negative,
(ii) at instant t_2, $i_2 = 0$, i_3 is positive and i_1 is negative,
(iii) at instant t_3, $i_3 = 0$, i_1 is positive and i_2 is negative.

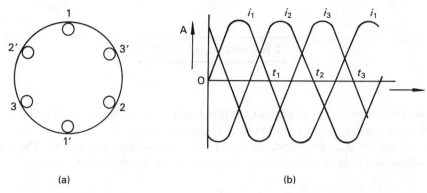

(a) (b)

The resultant flux due to these three currents at these instants is shown in the following three diagrams, from which it can be seen that the resulting flux axis is rotating in a clockwise direction.

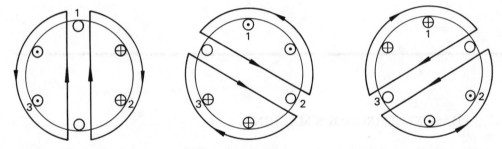

The currents shown have a 'phase sequence' 123, i.e. they reach their maximum value in this order. The phase sequences can be reversed so that the currents reach their maximum value in the order 321 by reversing the connections to any two of the coils.

What would be the effect of reversing the phase sequence?

58

The magnetic field would rotate in the opposite direction

After a complete cycle of current, when i_1 is zero again, the field has rotated through one revolution. If the currents are alternating at a frequency f Hertz and the time for one cycle is T second, then $T = 1/f$.

The speed of a 2-pole field is thus 1 revolution in $1/f$ second $= f\,(r/s)$. If we consider a 4-pole field there will be six coils and twelve coil sides and the two sides of a coil will be 90° apart.

What will be the speed of rotation of this 4-pole field?

59

$f/2$ revolutions per second

In this case, after one cycle of current, the flux axis has moved through half a revolution which is equivalent to $f/2$ revolutions per second.

In general, for a $2p$ pole field, the speed is f/p revolutions per second. The speed of rotation of a field is called the synchronous speed. N_s, so that in general

$$N_s = \frac{f}{p} \; (\text{r s}^{-1})$$

60

3-PHASE SYNCHRONOUS MACHINES

There are two main parts to this type of machine: the stator and the rotor. The stator carries a 3-phase winding called the armature winding located in slots around the inner periphery of a laminated iron core.

The rotor carries a d.c. winding called the field winding which is fed via slip rings mounted on the shaft of the machine. The rotor may be cylindrical in shape (in the case of high speed machines) or it may have salient poles. Stationary North and South poles are produced on the rotor by the current in the field winding. The diagram shows these two types of rotor.

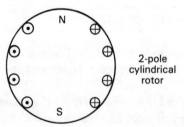

2-pole
cylindrical
rotor

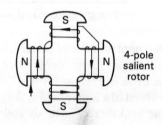

4-pole
salient
rotor

SYNCHRONOUS GENERATOR OPERATION

In this mode of operation the rotor is driven at synchronous speed by an external prime mover. A cylindrical rotor would be driven by a steam turbine, for example, whereas a salient pole rotor would be driven at slower speeds by, for example, a diesel engine. The stationary magnetic field produced by the field winding thus rotates at synchronous speed and cuts the armature conductors. Three-phase e.m.f.s are thus induced in the 3-phase winding and these are available for supplying external circuits.

If the rotor has 2 poles and is rotated at 3000 r/min, then the frequency of the e.m.f. generated $f = pN_s = 1 \times 3000/60 = 50$ Hz.

At what speed must a 2-pole machine be driven in the USA where the generated voltage is at a frequency of 60 Hz?

$$\boxed{3600 \text{ r/min}}$$

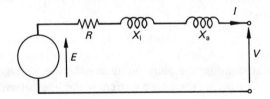

A 3-phase synchronous generator can be represented by the 1-phase equivalent circuit shown in which E and V are respectively the generated e.m.f. and the terminal voltage per phase; R and X_1 are respectively the armature resistance and leakage reactance per phase; X_a is called the armature reaction reactance per phase and takes account of armature reaction effects. The two reactances X_1 and X_a are usually combined into one reactance called the synchronous reactance per phase, symbol X_s.

Write down the Kirchhoff voltage equation for the equivalent circuit shown.

$$V = E - IR - IX_1 - IX_a = E - IR - IX_s \ldots \textit{phasorially}$$

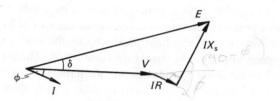

The phasor diagram shown above is for a lagging load current.

The voltage drop in the armature resistance is relatively small, about 1% of the terminal voltage V, and is often neglected in calculations, particularly in the case of the large machines found in power stations. The voltage drop in the armature winding leakage reactance X_1 is of the order of 10% of the terminal voltage V. However, whereas the effects of armature reaction in d.c. machines are not very great, in synchronous machines they are extremely important. It is as if the machine had a large inductive reactance X_a and the voltage drop in this inductance on full load could be as much as twice or even more times as big as the terminal voltage V.

Example

A 3-phase, 6600 V star connected synchronous generator has a resistance of 0.5 Ω per phase and a synchronous reactance of 5 Ω per phase,.

Determine the generated e.m.f. per phase required to supply a load current of 120 A at unity power factor.

Solution

For a star connected winding, the phase voltage = the line voltage$/\sqrt{3}$,
the phase current = the line current.

In this case

$$V = 6600/\sqrt{3} = 3810 \text{ V}$$

$$IR = 120 \times 0.5 = 60 \text{ V}$$

$$IX_s = 120 \times 5 = 600 \text{ V}$$

$$\cos \phi = 1 \text{ so that } \phi = 0°$$

therefore

$$E^2 = (V + IR)^2 + (IX_s)^2 = 3870^2 + 600^2$$

and

$$\underline{E = 3916 \text{ V}}$$

64

SYNCHRONOUS MOTOR OPERATION

In this mode of operation, 3-phase currents are fed into the armature winding and a rotating magnetic field is set up as discussed in Frame 57. This field sweeps past the rotor poles at high speed so that they experience an alternating torque urging them first in one direction then the other. The result is that they remain stationary. In order to start a synchronous motor, therefore, it is necessary to run it up to near synchronous speed by some means (for example, another motor) and then, when the field excitation is switched on, the rotor will pull into synchronism. Both the armature and rotor fields are then rotating at the same speed so that a unidirectional torque is produced, provided that there is an angle between their axes. The greater the load on the motor, the greater will be the torque required and the larger will be the angle between the axes of the two fields. This angle is in fact called the load angle (symbol δ) of the machine; it changes in magnitude whenever the load on the machine changes and it changes in sign whenever the machine changes mode of operation (motoring to generating or vice versa). For any given load, however, δ is constant and in the steady state the machine runs at constant (synchronous) speed.

The equivalent circuit for a motor is similar to that for a generator but now the terminal voltage, V, is the input voltage and the current is reversed.

Sketch this equivalent circuit and write down the corresponding Kirchhoff voltage equation.

65

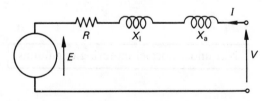

$$V = E + IR + IX_1 + IX_a \qquad \text{a phasor sum}$$

66

3-PHASE INDUCTION MACHINES

The vast majority of induction machines operate in the motoring mode. Like all rotating machines, they consist of a stator and a rotor. The stator is made in a similar way to that of the synchronous machine and it carries a similar 3-phase winding called the stator winding.

There are two types of rotor, namely the squirrel cage, and the wound rotor. The squirrel cage rotor consists of solid copper bars, located in slots in a laminated core, and whose ends are short-circuited by a copper or alloy end ring. This type is cheap and robust but its electrical characteristics are fixed. The wound rotor consists of a 3-phase winding similar to that of the stator, usually connected in star with its ends brought out to slip rings mounted on the shaft. The winding is short-circuited when the machine is running normally, but extra resistance can be inserted via the slip rings for starting and speed-control purposes.

MOTORING OPERATION

When 3-phase currents are supplied to the stator winding, a rotating magnetic field is set up which travels at synchronous speed $N_s = f/p$ where f is the frequency of the supply currents and p is the number of machine pole pairs. This rotating field cuts the rotor conductors and induces e.m.f. s in them which circulate current around the short-circuited winding. The direction of these e.m.f.s is, according to Lenz's law, such as to oppose their being set up, and the only way for this to happen is for there to be no relative motion between the rotating magnetic field and the rotor conductors. The rotor sets off after the field, therefore, trying to catch it up.

Can the rotor ever catch up with the field?

67

Not under normal motoring conditions

If the rotor caught up with the field there would be no induced e.m.f. in it, no current and no torque. A torque is required even on no load to overcome friction and other losses. The only way in which synchronous speed can be reached is if the machine is driven there by an external means or if the load 'runs away' with the motor as a result of gravitational force.

68

SLIP

The difference between the speed of the rotor (call it N_r) and the speed of the rotating magnetic field (N_s) is called the slip speed (N_{sr}). The fractional or per unit slip (symbol s) is defined as N_{sr}/N_s.

$$\therefore \quad s = \frac{N_s - N_r}{N_s} \Rightarrow sN_s = N_s - N_r \quad \text{and} \quad \underline{N_r = (1-s)N_s}$$

When the motor is at rest, $N_r = 0$ and $s = 1$.
If the motor could run at synchronous speed, $N_r = N_s$ and $s = 0$.
Typical values of full load slip lie between 0.02 and 0.06.
   What is the slip of a 4-pole motor running at a speed of 1425 r/min when it is connected to a 50 Hz supply?

69

$$\boxed{0.05 \text{ p.u.}}$$

Here is the working:

Synchronous speed, $N_s = f/p = 50/2 \text{ r s}^{-1} = 1500 \text{ r/min}$.

Slip $\qquad\qquad s = \frac{N_s - N_r}{N_s} = \frac{1500 - 1425}{1500} = 0.05 \text{ p.u.}$

The frequency of the e.m.f.s induced in the rotor (f_r, say) is dependent on the relative speed of the rotating magnetic field and the rotor itself.

This is $\qquad\qquad N_{sr} = N_s - N_r$

Just as $\qquad\qquad f = N_s p$, so $f_r = N_{sr} p = (N_s - N_r)p$

But $\qquad\qquad N_s - N_r = sN_s$

$$\therefore \quad f_r = sN_s p = sf$$

   What is the frequency of the rotor e.m.f.s of a 4-pole 50 Hz motor (a) at standstill and (b) when running at 1425 r/min?

$$\boxed{\text{(a) 50 Hz, (b) 2.5 Hz}}$$

LOSSES AND EFFICIENCY

The way in which the power input to a 3-phase induction motor is distributed is shown in the following power flowchart

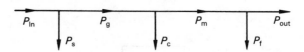

In this chart P_{in} is the total input power to the motor

P_s is the stator iron and copper loss

P_g is the input power to the rotor via the magnetic field

P_c is the rotor copper loss

P_m is the gross mechanical power developed by the rotor

P_f is the friction and windage loss

P_{out} is the useful mechanical power available at the shaft.

At normal speeds, the rotor frequency is very small so that the iron losses in the rotor are negligible.

The power input to the rotor is equal to the power leaving the stator i.e. $P_g = 2\pi N_s T$ (W) where T is the torque acting (N m).

The mechanical power developed, $P_m = 2\pi N_r T \text{ (W)} = 2\pi(1-s)N_s T \text{ (W)}$

$$\text{It follows that } P_m = (1-s)P_g \tag{i}$$

From the chart we have that $\quad P_g = P_c + P_m$

$$= P_c + (1-s)P_g$$

$$\therefore \quad P_c = sP_g \tag{ii}$$

Using (i) and (ii), obtain a general expression relating P_g, P_m, P_c and s.

$$\boxed{P_g: P_m: P_c :: 1 : (1-s) : s}$$

This is a shorthand way of writing

$$\frac{P_g}{1} = \frac{P_m}{1-s} = \frac{P_c}{s} \quad \text{and} \quad \frac{1}{P_g} = \frac{1-s}{P_m} = \frac{s}{P_c}.$$

Knowing any one of P_g, P_m and P_c together with the slip s, we can thus calculate the other two.

From the power flowchart we also have that $P_{in} = P_s + P_g$ and $P_m = P_f + P_{out}$.

Example

A 3-phase, 4-pole induction motor has a stator loss of 1 kW and a friction and windage loss amounting to 5% of the gross mechanical power developed.
Determine (a) the slip,

              (b) the rotor copper loss,

and   (c) the efficiency when the motor runs at 1455 r/min and takes a total power of 60 kW from the supply.

Solution

(a) The slip $s = \dfrac{N_s - N_r}{N_s}$ p.u.

    $N_s = 60f/p = 3000/2 = 1500$ r/min

    $N_r = 1455$ r/min

    $\therefore\quad s = (1500 - 1455)/1500 = \underline{0.03}$ p.u.

(b) Using the power flowchart given in Frame 70
    we have that $P_g = P_{in} - P_s$

        $\therefore\quad P_g = 60 - 1 = 59$ kW

From the expressions derived above:

        $P_c = sP_g$ and in this case $s = 0.03$ p.u.

    $\therefore\quad P_c = 0.03 \times 59 = \underline{1.77 \text{ kW}}$

(c) We have that $P_m = P_g - P_c = 59 - 1.77 \quad = 57.23$ kW

    We are told that $P_f = 0.05 P_m = 0.05 \times 57.23 = 2.86$ kW

    The output power $P_{out} = P_m - P_f = 57.23 - 2.86 = 54.37$ kW

    The efficiency $= \dfrac{\text{output power}}{\text{input power}} = \dfrac{54.37 \text{ kW}}{60 \text{ kW}} = \underline{0.91 \text{ p.u.}}$

72

THE TORQUE OF A 3-PHASE INDUCTION MOTOR

The power leaving the stator is $2\pi N_s T$ (W), where T is the torque acting, and this must equal the power input to the rotor, P_g (the 'air gap power').

Thus

$$T = \frac{P_g}{2\pi N_s} \text{ (N m)}$$

Now

$$P_g = \frac{P_c}{s} \quad \text{so that} \quad T = \frac{P_c}{s2\pi N_s} = \frac{3I_2^2 R_2}{s2\pi N_s} \text{ (N m)}$$

P_c is the total copper loss in the three phases and is given by $3I_2^2 R_2$ (W) where I_2 and R_2 are the rotor current and resistance per phase respectively.

The rotor e.m.f. per phase at standstill is E_2 and is due to the relative speed (N_{sr}) between the rotor and the rotating magnetic field. At a slip s the relative speed is reduced to $N_s - N_r = sN_s$ so that the e.m.f. induced in the rotor will be s times the standstill value, i.e. sE_2 volts per phase.

The rotor impedance per phase at standstill is given by $Z_2 = (R_2^2 + X_2^2)^{1/2}$ where X_2 is the rotor inductive reactance per phase at standstill and is $2\pi f$ times the rotor inductance per phase, i.e. $X_2 = 2\pi f L_2$.

What will be the rotor impedance when the motor is running with a slip s?

73

$$\boxed{(R_2^2 + (sX_2)^2)^{1/2}}$$

At a slip s, the frequency in the rotor is $f_r = sf$ so the inductive reactance is s times its standstill value (when the frequency is f). The rotor current at a slip s is given by

$$\frac{sE_2}{Z_2} = \frac{sE_2}{(R_2^2 + (sX_2)^2)^{1/2}}$$

Substituting this into the torque expression,

$$T = \frac{3(sE_2)^2 R_2}{sZ_2^2 2\pi N_s} \text{ (N m)}$$

From which

$$T = \frac{3sE_2^2 R_2}{R_2^2 + (sX_2)^2} \frac{1}{2\pi N_s} \text{ (N m)} \tag{i}$$

In this expression, the only variable is s and if T is plotted to a base of s the following characteristic is obtained:

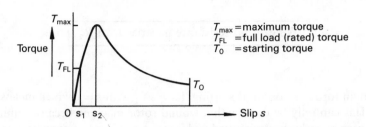

T_{max} = maximum torque
T_{FL} = full load (rated) torque
T_0 = starting torque

For a motor of normal design, s_2 is typically 0.2 and s_1 is 0.02–0.06.

Is this speed characteristic similar to that of a d.c. series motor or to that of a d.c. shunt motor?

74

| The 3-phase induction motor has a 'shunt' characteristic |

Note that as the torque varies from 0 to T_{fl}, so the slip rises from 0 to 0.05 (approximately) and the speed falls from N_s to $(1 - 0.05)N_s = 0.95N_s$ (approximately), i.e. there is a slight fall from no load to full load.

By differentiating the expression for torque with respect to s and equating to zero we can obtain the condition for maximum torque.

$\dfrac{3E_2^2 R_2}{2\pi N_s}$ is a constant, k say, so that $T = \dfrac{ks}{R_2^2 + (sX_2)^2} = \dfrac{u}{v}$ say

$$\frac{dT}{ds} = \frac{u(dv/ds) - v(du/ds)}{v^2} = \frac{ks(2sX_2^2) - (R_2^2 + (sX_2)^2)k}{v^2}$$

Putting this equal to zero, we get $2k(sX_2)^2 = kR_2^2 + k(sX_2)^2$

$$\therefore \quad sX_2 = R_2$$

Differentiating a second time gives a negative value (i.e. $d^2 T/ds^2$ is negative), indicating a maximum value for torque. For maximum torque, therefore, the rotor resistance must equal the rotor reactance. Machines of normal design have $X_2 \approx 5R_2$ so that $s \approx 1/5 = 0.2$ p.u. for maximum torque (T_{max}).

What can be done to obtain the maximum torque at starting, when $s = 1$?

75

> We can put resistance in series with R_2

For maximum torque to occur at starting, $R_2 = sX_2$ with $s = 1$ which means making $R_2 = X_2$. This can only be done with a wound rotor machine because squirrel cage rotors are permanently short-circuited and it is not possible to put resistance in series with them.

Obtain an expression for the maximum torque of a 3-phase induction motor.

76

$$T_{max} = \frac{3E_2^2}{4\pi N_s} \ (\text{N m})$$

This is obtained by putting $s = R_2/X_2$ in the torque expression ((i) Frame 73).

Note that the expression for maximum torque is a constant for a given motor and is independent of rotor resistance. Adding resistance to the rotor circuit simply alters the slip and therefore the speed at which the maximum torque occurs. It does not alter the *magnitude* of the maximum torque.

Example

A 3-phase wound rotor induction motor has six poles and is connected to a 50 Hz supply. The rotor resistance is 0.1 Ω per phase and the inductive reactance at standstill is 0.5 Ω per phase.

Determine the speed at which the maximum torque occurs.

Solution

The synchronous speed is $N_s = f/p = (50 \times 60)/3 = 1000$ r/min.
For maximum torque, $s = R_2/X_2 = 0.1/0.5 = 0.2$ p.u.
The corresponding speed is $N_r = (1 - s)N_s = (1 - 0.2)1000 = \underline{800 \text{ r/min}}$.

77

INDUCTION GENERATOR

If an induction motor is running on no load its slip is very small and the rotor-induced e.m.f., current and torque are all small positive values. If, now, an external prime mover is used to drive the machine so that it runs at synchronous speed, the induced e.m.f., current and torque will all become zero. Suppose that the prime mover speed is increased so that the induction machine is driven above synchronous speed. The slip will be < 0 (i.e. negative) and an e.m.f. will reappear in the reverse direction which will produce a negative current. This will give rise to negative torque which acts against the direction of rotation and, if the machine is to retain its supersynchronous speed, the prime mover torque must be greater than this. The machine thus receives power from the prime mover and supplies it to the electricity supply to which it is connected.

The induction generator is not self-exciting and receives its excitation from the supply into which it feeds energy. Its applications are somewhat limited, one being for making use of small, variable amounts of water by using a water-wheel to drive the machine. When there is sufficient water to drive it above synchronous speed, the machine is switched in to an existing system and delivers power to it.

78

SINGLE-PHASE a.c. MOTORS

(a) Single-phase series motor: this type of motor has a commutator and is similar to a d.c. series motor except that its pole and yoke are laminated. The fluxes produced by the current flowing in the series field winding and the armature winding interact to produce the required torque. It is only made in small sizes (up to a few hundred watts), is used for driving small domestic equipment, operates on a.c. or d.c., and is called a universal motor.

(b) Single-phase induction motor: this type of motor is made in sizes less than about 5 kW and the vast majority are less than 1 kW. It is relatively inefficient and operates at a low power factor. The alternating magnetic field produced by single-phase currents are not rotating and so the single-phase motor has no starting torque. However, it can be shown that the alternating field may be resolved into two fields rotating in opposite directions so that, if the motor is started by hand or by some other means in either direction, it will run up to speed in that direction. One way of achieving this is by 'phase splitting' which involves using a starting winding in conjunction with a resistor, inductor or capacitor. To illustrate the principle, a capacitor start motor is described in the next frame.

79

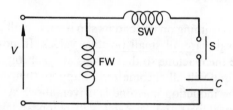

The capacitor ensures that the currents through the main field winding (FW) and the starting winding (SW) are out of phase so that there is an effective 2-phase supply to the motor. These out-of-phase currents produce a rotating magnetic field in much the same way as described in Frame 57 so that there is a unidirectional torque produced and the motor will start. Once up to speed, a centrifugal switch (S) isolates the auxiliary starting equipment.

80

Now decide which of the following statements are true.

(a) 3-phase induction machines rely for their operation on the production of an alternating magnetic field.

(b) Synchronous machines are wound with their armature winding on the stator.

(c) $N_s = f/p$ where the symbols have their usual meaning.

(d) The field winding of a synchronous machine is fed with d.c.

(e) Armature reaction effects in a synchronous machine are less important than those in a d.c. machine.

(f) The effect of armature reaction in a synchronous machine is as if it had a reactance in which a very large voltage drop occurs on full load.

(g) Synchronous motors are not self-starting.

(h) A 3-phase induction motor runs at full load with a slip of about 0.2 p.u.

(i) A squirrel cage rotor is permanently short-circuited.

(j) The frequency of the rotor currents f_r = slip × the supply frequency.

(k) At synchronous speed, $s = 1$.

(l) $P_g/P_m = 1/(1 - s)$.

(m) The torque/speed characteristic of a 3-phase induction motor is similar to that of a d.c. series motor.

(n) When an induction machine is operating as a generator, the slip is > 1.

(o) Single-phase induction motors have no inherent starting torque.

81

True: b, c, d, f, g, i, j, l, o

(a) They rely on a rotating magnetic field. (Frame 57)
(e) It is much more important in the synchronous machines. (Frame 63)
(h) Full load slip is normally about 0.05 p.u. (Frame 68)
(k) At synchronous speed $s = 0$. (Frame 68)
(m) The 3-phase induction motor has a shunt characteristic. (Frame 74)
(n) When generating, the slip of an induction machine is negative. (Frame 77)

82

SUMMARY OF FRAMES 57–81

$$N_s = \frac{f}{p} \, (\text{r s}^{-1})$$

the synchronous speed of a rotating magnetic field where p is the number of pole *pairs*

$$V = E \mp IZ_s$$

the Kirchhoff voltage equation applied to the per phase equivalent circuit of a synchronous generator $(-)$ or motor $(+)$

$$Z_s^2 = R^2 + X_s^2$$

the synchronous impedance of a synchronous machine (per phase) in terms of its armature resistance per phase and the synchronous reactance per phase

$$N_r = (1-s)N_s$$

the rotor speed of a 3-phase induction motor in terms of the slip and the synchronous speed

$$f_r = sf$$

the frequency of the rotor e.m.f.s and currents in terms of the slip and the supply frequency

$$P_g : P_m : P_c :: 1 : (1-s) : s \quad \text{rotor power relationships}$$

$$T = \frac{3sE_2^2 R_2}{R_2^2 + (sX_2)^2} \times \frac{1}{2\pi N_s} \, (\text{N m}) \quad \text{induction motor torque equation}$$

For maximum torque, $R_2 = sX_2$

83

EXERCISES

(i)   At what speed must a 4-pole synchronous generator run in order to obtain a frequency of (a) 50 Hz, (b) 60 Hz, (c) 400 Hz?

(ii) A 6.6 kV, star connected, 3-phase synchronous machine has a resistance of 2 Ω per phase and a synchronous reactance of 20 Ω per phase.

Calculate the generated e.m.f. per phase when the machine operates: (a) as a generator delivering a current of 330 A at a power factor of 0.8 lagging and (b) as a motor taking a current of 200 A at unity power factor.

(iii) The input power to a 3-phase, 6-pole, 50 Hz induction motor is 100 kW when it operates with a slip of 0.03 p.u. Friction and windage losses total 1000 W and stator losses are 750 W.

Calculate the rotor copper loss per phase and the efficiency.

(iv) A 3-phase, 4-pole, 50 Hz induction motor has a rotor resistance per phase of 0.02 Ω and a rotor reactance per phase of 0.1 Ω.

Determine the speed at which maximum torque is developed and the amount of resistance per phase which must be placed in series with the rotor winding in order to obtain maximum torque at starting.

84

SHORT EXERCISES ON PROGRAMME 9

The number in brackets refers to the frame where the answer is to be found. All symbols have the meaning given in the programme.

1. Give an expression for the e.m.f. generated in a d.c. armature.   (7)
2. Give an expression for the torque developed in a d.c. armature.   (9)
3. What is meant by commutation in d.c. machines?   (10)
4. Draw the equivalent circuit of a d.c. shunt motor.   (17)
5. Sketch the torque characteristic of a series motor.   (21)
6. How may the speed of d.c. motor be varied?   (25)
7. Why does a d.c. shunt motor need a 'starter'?   (37)
8. What is meant by 'level compounding' of d.c. generators?   (46)
9. What is the effect of armature reaction in d.c. machines?   (47)
10. Give the condition for the maximum efficiency of a d.c. machine.   (50)
11. Give the relationship between N_s, f and p.   (59)
12. Sketch the equivalent circuit for a synchronous generator.   (62)
13. Define fractional slip.   (68)

14. What is the relationship between f_r, s and f? (69)
15. Draw the power flowchart for a 3-phase induction motor. (70)
16. Give the relationship between P_g, P_m, P_c and s. (71)
17. Sketch the torque–slip curve for a 3-phase induction motor. (73)
18. What is the condition for maximum torque of a 3-phase induction motor? (74)
19. How can a 3-phase induction machine become a generator? (77)
20. Name two types of 1-phase a.c. motor. (78)

85

ANSWERS TO EXERCISES

Frame 13

(i) (a) 282 V, (b) 80.79 N m;
(ii) 30.49 mWb;
(iii) 318.75 V, 30 A.

Frame 41

(i) 2 A, 188.8 V;
(ii) 5.3 Ω;
(iii) 20 Ω;
(iv) (a) 637 r/min, (b) 683 r/min.

Frame 56

(i) (a) 173 V, (b) 45 Ω;
(ii) 7.5 (say 8) series field turns per pole;
(iii) (a) 0.891 p.u., (b) 0.892 p.u. when the armature current is 126.5 A.

Frame 83

(i) (a) 25 r/s (1500 r/min), (b) 30 r/s (1800 r/min), (c) 200 r/s (12000 r/min);
(ii) (a) 9.628 kV, (b) 5.256 kV;
(iii) 992.5 W, 0.9527 p.u.;
(iv) 20 r/s (1200 r/min), 0.08 Ω per phase.